AVR RISC Microcontroller Handbook

AVR RISC Microcontroller Handbook

by Claus Kühnel

Newnes

Boston Oxford Johannesburg Melbourne New Delhi Singapore

Newnes is an imprint of Butterworth–Heinemann.

Copyright © 1998 by Butterworth–Heinemann

 A member of the Reed Elsevier group

 Recognizing the importance of preserving what has been written, Butterworth–Heinemann prints its books on acid-free paper whenever possible.

 Butterworth–Heinemann supports the efforts of American Forests and the Global ReLeaf program in its campaign for the betterment of trees, forests, and our environment.

Library of Congress Cataloging-in-Publication Data
Kühnel, Claus, 1951–
    AVR RISC microcontroller handbook / by Claus Kühnel.
      p.   cm.
    Includes index.
    ISBN 0-7506-9963-9 (alk. paper)
     1. Programmable controllers.   2. RISC microprocessors. I. Title.
  TJ223.P76K83   1998
629.8′9—dc21   98-14859
                                               CIP

British Library Cataloguing-in-Publication Data
A catalogue record for this book is available from the British Library.

The publisher offers special discounts on bulk orders of this book.
For information, please contact:
Manager of Special Sales
Butterworth-Heinemann
225 Wildwood Avenue
Woburn, MA 01801-2041
Tel: 781-904-2500
Fax: 781-904-2620

For information on all Butterworth–Heinemann publications available, contact our World Wide Web home page at: http://www.bh.com

10 9 8 7 6 5 4 3 2 1

Transferred to digital printing 2006

Contents

Preface

The AVR enhanced microcontrollers are based on a new RISC architecture that has been developed to take advantage of semiconductor integration and software capabilities in the 1990s. The resulting microcontrollers offer the highest MIPS/mW capability available in the 8-bit microcontroller market.

High-level languages are rapidly becoming the standard programming methodology for embedded microcontrollers because of improved time-to-market and simplified maintenance support. The AVR architecture was developed together with C language experts to ensure that the hardware and software work hand-in-hand to develop highly efficient, high-performance code.

To optimize code size, performance, and power consumption, the AVR architecture has incorporated a large fast-access register file and fast single-cycle instructions.

The AVR architecture supports a complete spectrum of price performance from simple small-pin-count controllers to high-range devices with large on-chip memories. The Harvard-style architecture directly addresses up to 8 Mbytes of data memory. The register file is dual mapped and can be addressed as a part of the on-chip SRAM memory to enable fast context switching.

The AVR enhanced RISC microcontroller family is manufactured with ATMEL's low-power nonvolatile CMOS technology. The on-chip in-system-programmable (ISP) downloadable flash memory allows the program memory to be reprogrammed in-system through an SPI serial port or conventional memory programmer. By combining an enhanced RISC architecture with downloadable flash memory on the same chip, the AVR microcontroller family offers a powerful solution to embedded control application.

This book describes the new AVR architecture and the program development for those microcontrollers of the family available in early 1997. Some tools from ATMEL and third-party companies help to give a first impression of the AVR microcontrollers. Thus, the evaluation of hardware and programming in Assembler and C of that type of microcontroller is supported very well. A simulator makes program verification possible without any hardware.

The development of the AVR microcontroller family by ATMEL shows clearly that remarkable results are not limited to high-end microcontrollers that are often the focus of consideration.

I thank ATMEL for the development of this interesting microcontroller family, because studying these new devices and their development environment was very interesting and made writing this book enjoyable.

Finally, I wish to thank ATMEL Norway and IAR Sweden for their support of this project and my wife, Jutta, for her continued understanding during the preparation of this book.

Some Basics 1

In the following chapters we will use some special terms that are perhaps not so familiar to a beginner. Some explanations should make the world of micro-controller terms and functionality more transparent.

1.1 Architecture

All microcontrollers have more or less the same function groups. Internally we find memory for instructions and data and a central processing unit (CPU) for handling the program flow, manipulating the data, and controlling the peripherals.

Figure 1-1 shows the basic function blocks in a microcontroller. The CPU communicates with each of these function blocks (memory and peripherals).

To build a powerful microcontroller, it is important to reduce the tasks carried out from the CPU itself and to optimize the handling of the remaining tasks.

On the left side of Figure 1-1, peripherals are arranged. These peripherals react with the world outside, or in more technical terms, with the process. In modern microcontrollers, the peripherals relieve the CPU by handling the external events separately.

In an asynchronous serial communication, for example, the CPU transmits the character to be sent only to the serial port. The required serialization and synchronization are performed by the serial port itself. On the other side, re-ceiving a character is important for the CPU only when all bits are stored in a buffer and are ready for an access of the CPU.

A port is built by a certain number of connections between the microcon-troller and the process, often a factor of 8. It usually supports a bitwise digital input and/or output (I/O).

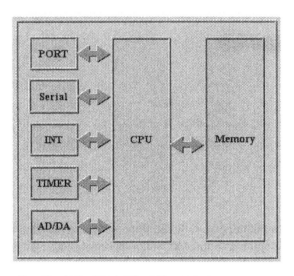

Figure 1-1
Microcontroller function blocks.

Serial ports communicate with other external components by means of serial communication protocols. Asynchronous and synchronous serial communications must be differentiated. Both kinds of serial communication have their own building blocks—Universal Asynchronous Receiver and Transmitter (UART) for asynchronous communication, and Serial Peripheral Interface (SPI) for synchronous communication. In Figure 1-1 this differentiation is not shown.

Because the microcontroller is designed for process-related applications with real-time character, some other function groups are implemented in a microcontroller.

Modern microcontrollers have a fairly comfortable interrupt system. An interrupt breaks the running program to process a special routine, called the interrupt service routine. Some external events require an immediate reaction. These events generate an interrupt; the whole system will be frozen, and in an interrupt service routine the concerning event is handled before the program goes on. If the microcontroller has to process many interrupts, an interrupt controller can be helpful.

To fulfill timing conditions, microcontrollers have one or more timers implemented. The function blocks usually work as timer and/or counter subsystems. In the simplest case, we have an 8-bit timer register. Its contents will be incremented (or decremented) with each clock cycle (CLK). Any time the value 255 (or 0) is reached, the register will overflow (or underflow) with the next clock. This overflow (or underflow) is signalized to the CPU. The actual delay depends on the preload value.

If the preload value equals zero, then the overflow will occur after 256 clock periods. If the preload value equals 250, in an up-counter the overflow will occur after six clock cycles. A block diagram of a simple timer is shown in Figure 1-2.

Last but not least, we had in Figure 1-1 an AD/DA subsystem. For adaptions to a real-world process, analog-to-digital and/or digital-to-analog converters are often required. This is not the place to discuss all features of these more or less complex subsystems. Some microcontrollers include such an AD/DA subsystem or can control an external one. In other cases, only analog comparators are integrated.

Thus, each peripheral has its own intelligence for handling external events and to realize preprocessing.

On the right side of Figure 1-1, we find the microprocessor part (CPU and memory). The memory contains program and data. CPU and memory are connected through a bus system. This architecture—called "von Neumann" architecture—has some drawbacks.

In Figure 1-3 instruction handling is explained. Before an instruction can be operated, it must first be fetched from memory. Next, the program counter must be incremented. After this incrementation, the program counter points to the next instruction in memory. Before execution of the fetched instruction, it has to be decoded. As a result of this decoding, further memory accesses for operands or addresses are possible. The instruction execution includes arithmetic or logical operations, followed by storage of the result back to the memory.

It seems not so difficult to understand that the handling of one instruction requires more than one memory access. Usually, one instruction manipulates one data byte. Therefore, several memory accesses are inconsistent with the manipulation of one data byte by one instruction.

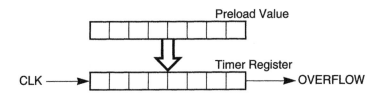

Figure 1-2
Timer.

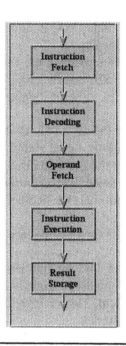

Figure 1-3
Instruction handling.

Very large scale integration (VLSI) technology has made it possible to build very fast CPUs. However, slow memory access times inhibits this evolution. This fundamental problem of the von Neumann architecture is called the "von Neumann bottleneck."

To avoid these limitations, system designers implement some sort of cache memory—a fast memory buffer between the main memory and the CPU. Another, and most recent approach, is to separate the paths to the memory system for instructions and data—the "Harvard" architecture.

The RISC design philosophy also tries to eliminate the "von Neumann bottleneck" by strict limitation of memory operations by means of many internal registers.

In order to increase performance, pipelining is widely used in modern microprocessor architectures. A basic linear pipeline consists of a cascade of processing stages.

The instruction unit in a microprocessor architecture can consist of pipeline stages for instruction fetch, instruction decode, operand fetch, execute instruction, and store results. Using such a technique for preparing an execute operation allows a new instruction to be executed every clock cycle. In this

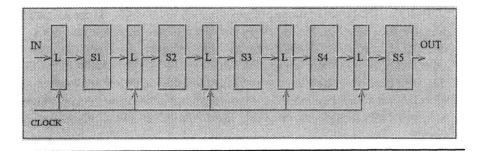

Figure 1-4
Instruction pipeline.

design, five different instructions are flowing simultaneously in the pipeline. Figure 1-4 shows an instruction pipeline for all steps of the instruction handling according to Figure 1-3.

All blocks in the pipeline are pure combinational circuits performing arithmetic or logic operations over the data stream flowing through the pipeline. The stages are separated by interface latches.

If a conditional program flow instruction changes the sequence in which the program is executed, the prefetched instructions, addresses, and operands in the instruction pipe are discarded. Different procedures are used to fill these empty slots with valid information. Since the branch-type instructions have damaging effects on the pipeline architecture performance, this is one of the most complex design stages in modern computer architecture.

1.2 Important Terms

The following important terms will help in understanding subsequent chapters.

- *Internal architecture*: The concept of building the internal electronics of a microcontroller.
- *Memory*: A function block for program and data storage. Here it is important to distinguish between nonvolatile and volatile memories. Nonvolatile memories are required for storage of programs so the system does not have to be reprogrammed after power has been off. Working variables and intermediate results need to be stored in a memory that can be written (or programmed) quickly and easily during system operation. It is not important to know these data after power off. Examples of nonvolatile memories are EPROM and

OTP-ROM. The typical example of volatile memory is RAM. There are many more types of memory, but here they are of no further interest in this book.

- *EPROM:* Electrical programmable read-only memory. This memory is programmed by programmer equipment. Memory is erased by irradiation of the chip with ultraviolet light through a crystal window in the ceramic package. EPROMs are typical memories for program storage. Some microcontrollers have EPROMs included so the CPU itself is involved in the programming process.

- *OTP-ROM*: One-time programmable EPROM. This type of EPROM is one-time programmable because the package is without a crystal window and therefore not UV erasable. Using cheap plastic packages without a window instead of windowed ceramic packages decreases the cost significantly. Therefore, the PROM included in a microcontroller is often an OTP-ROM.

- *EEPROM*: Electrical erasable and programmable read-only memory. For reading, it is no different from a normal EPROM. Writing or, better, programming an EEPROM differs from that for a normal EPROM completely. The program cycle is about 10 ms for a byte or a block of bytes. Because of the program time, the EEPROM is suitable for storage of seldom-changing data, such as initialization or configuration data. For modern EEPROMs, 10 million program cycles are possible.

- *Flash memory*: Nonvolatile read–write memory for program and data storage. Flash memories combine EPROM programming with *EEPROM*-like in-system electrical erasure. In contrast to EEPROM, a bytewise erasure is impossible.

- *RAM*: Random access memory, which can be programmed and read at any time. RAMs are typical memories for data storage and are volatile.

- *Oscillator*: A circuit that produces a constant-frequency square wave used by the computer as a timing or sequencing reference. A microcontroller typically includes all elements of this circuit except the frequency-determining component(s) (crystals, ceramic resonators, or RC components). In some cases all frequency-determining components are also on-chip.

- *Reset circuit*: Generates a reset impulse to reset the computer, in some cases. The most important reset is the power-on reset. Switching power-on starts the program of the microcontroller.

- *I/O ports*: The connections to the process. Such ports are mainly bit-programmable in both directions.

- *Watchdog:* A counter circuit that must be reset by the running program. If the program hangs, no watchdog reset can occur, and the watchdog counter overflows. As a result of this overflow, the watchdog initiates a reset and avoids wild running of the microcontroller

AVR RISC Microcontroller Handbook

- *Real-time clock/counter*: A further counter circuit able to count real-time pulses from the process side or controlled by a clock generated by the internal clock.
- *Terminal*: Equipment for serial I/O to the microcontroller. In most cases a PC running a terminal program is used.
- *Program installation*: Installation of an user program on the hard disk of a personal computer. The installation process includes copying the file(s), unencrypting these when needed, generating of a program group in a Windows environment, and so forth.
- *Program initialization*: To provide installed software with the required constants and/or parameters. For example, initialization would provide baud rate and handshake parameters for a serial communication.
- *Instruction set*: The whole list of instructions that will be understood by the microcontroller.
- *MSB*: Most significant bit. In the 8-bit data word D7:D0 = I0101010, the MSB is D7 = 1.
- *LSB*: Least significant bit. In the 8-bit data word D7:D0 = 1010101$0$, the LSB is D0 = 0.
- *Pull-up resistor*: A resistor that gives a Hi signal in a high-impedance circuit.
- *Pull-down resistor*: A resistor that gives a Lo signal in a high-impedance circuit.
- *Kbyte, Mbyte*: Units of bits and bytes. "K" here does not mean a value of 1000, and "M" does not mean 1,000,000. In the binary system used in information technology, "K" stands for $2^{10} = 1024$, and "M" for $2^{10} * 2^{10} = 1,048,576$.
- *PDIP*: Plastic dual inline package for integrated circuits.
- *SOIC*: Small-outline integrated circuit.
- *PLCC*: Plastic J-leaded chip carrier.

1.3 Numbers

Numbers can be displayed in various formats. Usually we think of decimal numbers. In digital systems, and also in the microcontroller world, we have to think binary because only the two states (Lo and Hi) are allowed.

For a byte-wide number (8 bits) we get the relations among binary, decimal, and hexadecimal number notation shown in Table 1-1.

To indicate the number system used, an index is usually used in text notation. The hexadecimal number 11_H is thus equivalent to the decimal 17_D. Normally, the index for decimal numbers is not shown.

Table 1-1

Notation of numbers.

Binary Number	Decimal Equivalent	Hexadecimal Equivalent	Notation in Text	Notation in Assembler or BASIC	Notation in C
0000 0000	0	0	0_H	$0000	0x0000
0000 0001	1	1	1_H	$0001	0x0001
0000 0010	2	2	2_H	$0002	0x0002
0000 0011	3	3	3_H	$0003	0x0003
. . .	. . .	. . .	. . .	. . .	. . .
0000 1110	14	E	E_H	$000E	0x000E
0000 1111	15	F	F_H	$000F	0x000F
0001 0000	16	10	10_H	$0010	0x0010
0001 0001	17	11	11_H	$0011	0x0011
. . . . . .	. . .	. . .	. . .	. . .	
1111 1110	126	FE	FE_H	$00FE	0x00FE
1111 1111	127	FF	FF_H	$00FF	0x00FF

In Assembler and BASIC, hexadecimal numbers are normally presented in the $-format. In C, the notation of a hexadecimal number is given in a special format. The rightmost column in Table 1-1 shows this format.

Hardware Resources of AVR Microcontrollers   *2*

As noticed in the preface, the AVR microcontroller family is based on a new RISC architecture. In order to optimize code size, performance, and power consumption, the AVR architecture has incorporated a large fast-access register file and fast single-cycle instructions. The fast-access RISC register file consists of 32 general-purpose working resisters. Traditional accumulator-based architectures require large amounts of program code for data transfers between the accumulator and memory. With these 32 working registers—each of which acts as an accumulator—in AVR microcontrollers, these data transfers are eliminated.

The AVR microcontroller prefetches an instruction during the previous instruction execution and then executes in a single cycle. In other CISC- and RISC-like architectures, the external oscillator clock is divided down (by as much as 12 times) to the traditional internal execution cycle. The AVR microcontrollers execute an instruction in a single clock cycle and are the first true RISC machines in the 8-bit market.

The AVR architecture supports a complete spectrum of price performance, from simple small-pin-count controllers such as the AT90S1200 on the low end, to high-range devices with large on-chip memories such as the AT90S8515

The Harvard-style architecture directly addresses up to 8 Mbytes of program memory and 8 Mbytes of data memory. The register file is dual mapped and can be addressed as part of the on-chip SRAM memory to enable fast context switching.

2.1 Architectural Overview

Two different CPU models are used in the AVR microcontroller family. Atmel's part numbering system gives the required information about the CPU model used in the each microcontroller device (see the Appendix for reference).

9

Figures 2-1 and 2-2 show block diagrams of the AT90S1200 (low end) and the AT90S8515 (high end) microcontrollers

Comparing Figures 2-1 and 2-2, we can find many common factors and some differences.

In all members of the AVR microcontroller family, the fast-access register file concept contains 32 8-bit general-purpose working registers with a single-clock-cycle access time. This means that during one single clock cycle, one

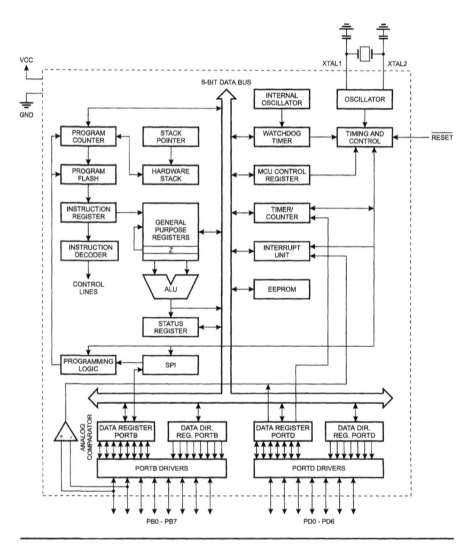

Figure 2-1

Block diagram of AT90S1200 microcontroller.

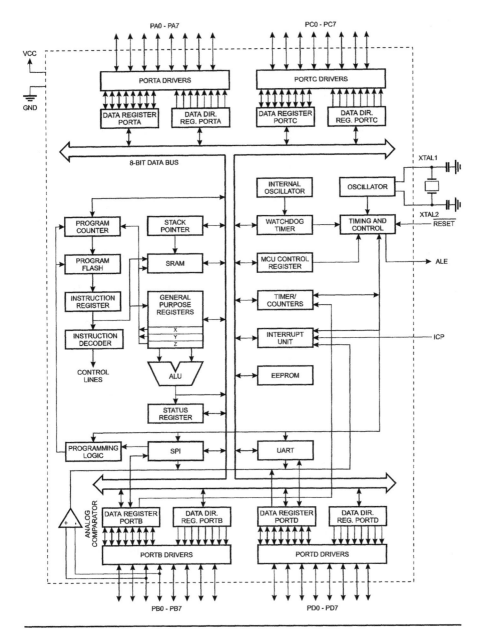

Figure 2-2
Block diagram of AT90S8515 microcontroller.

ALU (arithmetic logic unit) operation is executed. Two operands are fetched from the register file, the operation is executed, and the result is stored back in the register file—all in one clock cycle of 50 ns minimum.

In CPU model 1, six of the 32 registers can be used as three 16-bit indirect address register pointers for SRAM addressing, enabling efficient address calculations. In CPU model 0, only two registers build a 16-bit register. One of the three address pointers is also used as the address pointer for the constant table lookup function. These added function registers are the 16-bit X-register, Y-register, and Z-register.

In addition to the register operation, the conventional memory addressing modes can be used on the register file as well. This is enabled by the fact that the register file is assigned the 32 lowermost SRAM addresses, allowing them to be accessed as though they were ordinary memory locations.

With the relative jump and call instructions, the whole 4K address space is directly accessed. All AVR instructions have a single 16-bit word format, meaning that every program memory address contains a single 16-bit instruction.

During interrupts and subroutine calls, the return address program counter (PC) is stored on the stack. The stack is effectively allocated in the general data SRAM, and consequently the stack size is only limited by the total SRAM size and the usage of the SRAM. All user programs must initialize the stack pointer (SP) in the reset routine (before subroutines or interrupts are executed). The 16-bit SP is read/write accessible in the I/O space.

The 256-byte data SRAM can be easily accessed through the four different addressing modes supported in the AVR architecture.

The I/O memory space contains 64 addresses for CPU peripheral functions such as control registers, timer/counters, A/D converters, and other I/O functions. The memory spaces in the AVR architecture are all linear and regular memory maps.

The organization of program and data memory for an AT90S8515 microcontroller is shown in Figure 2-3 as a memory map. The program memory is executed with single-level pipelining. While one instruction is being executed, the next instruction is prefetched from the program memory. This concept enables instructions to be executed in every clock cycle. Figure 2-4 shows the timing conditions for this single-level pipelining.

A flexible interrupt module has its control registers in the I/O space with an additional global interrupt enable bit in the status register (SREG). All the different interrupts have a separate interrupt vector in the interrupt vector table at the beginning of the program memory. The different interrupts have priority in accordance with their interrupt vector position. The lower the interrupt address vector, the higher the priority.

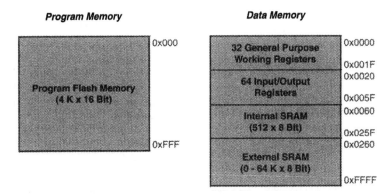

Figure 2-3
Memory maps for program and data memory for AT90S8515.

2.2 The Arithmetic Logic Unit

The high-performance AVR ALU operates in direct connection with all 32 general-purpose working registers. As Figure 2-5 shows, ALU operations between registers in the register file are executed within a single clock cycle.

The ALU operations are divided into three main categories—arithmetic, logic, and bit functions. Some microcontrollers in the AVR microcontroller family will feature a hardware multiplier in the arithmetic part of the ALU.

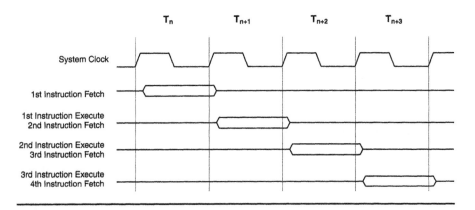

Figure 2-4
Parallel instruction fetch and execution.

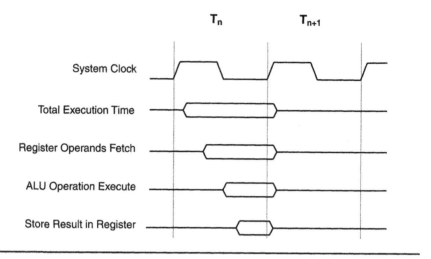

Figure 2-5
Single-cycle ALU operation.

2.3 Program and Data Memories

The AVR microcontroller uses a Harvard architecture concept, with separate memories and buses for program and data.

2.3.1 Downloadable Flash Program Memory

The program memory is in-system downloadable flash memory. The flash memory space of all microcontrollers of the AVR family is explained in detail in the hardware chapter. The different access methods to the program memory are explained in Chapter 3, which describes the handling of the hardware resources.

2.3.2 SRAM Data Memory

The data address space is organized in a maximum of four groups. The lower 96 bytes are reserved for register file and I/O registers in all devices of the AVR microcontroller family.

Microcontrollers with SRAM between internal and external SRAM will be distinguished. Figure 2-6 explains the organization of SRAM in an AT90S8515, for example. The parts of SRAM not available in all devices of the AVR microcontroller family are grayed. Microcontrollers have internal

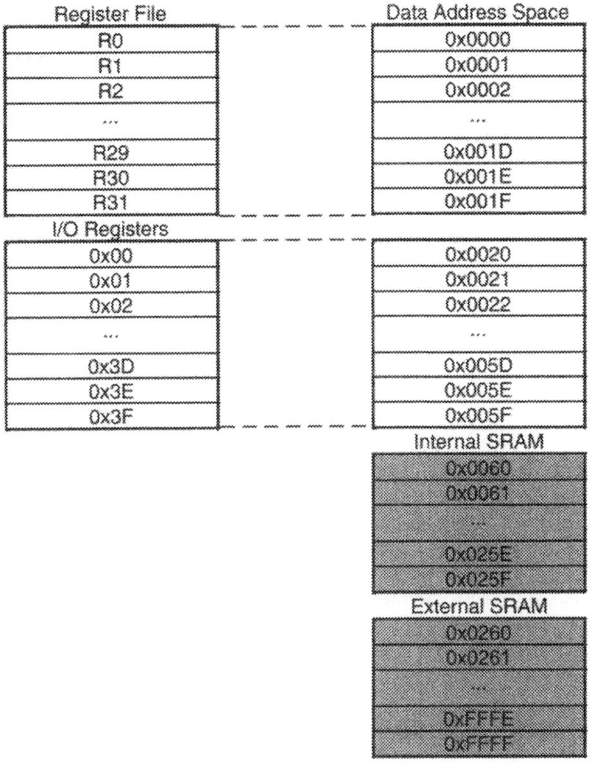

Register File		Data Address Space
R0		0x0000
R1		0x0001
R2		0x0002
...		...
R29		0x001D
R30		0x001E
R31		0x001F

I/O Registers		
0x00		0x0020
0x01		0x0021
0x02		0x0022
...		...
0x3D		0x005D
0x3E		0x005E
0x3F		0x005F

Internal SRAM
0x0060
0x0061
...
0x025E
0x025F

External SRAM
0x0260
0x0261
...
0xFFFE
0xFFFF

Figure 2-6
SRAM organization.

SRAM and can have external SRAM for enhancement. Between these both types must distinguished.

An access to the external SRAM occurs with the same instructions as for internal data SRAM access. When the internal data SRAM is accessed, the read and write strobe pins (/RD and /WR) are inactive during the whole access cycle. The external data SRAM physical address locations corresponding to the internal data SRAM addresses cannot be reached by the CPU. External SRAM operation is enabled by setting the SRE bit in the MCUCR register.

The PortA and PortC pins have alternative functions related to the optional external data SRAM. PortA can be configured to be the multiplexed low-order address/data bus during accesses to the external data memory, and PortC can be configured to be the high-order address byte.

Figure 2-7 shows read and write access to external SRAM without wait states.

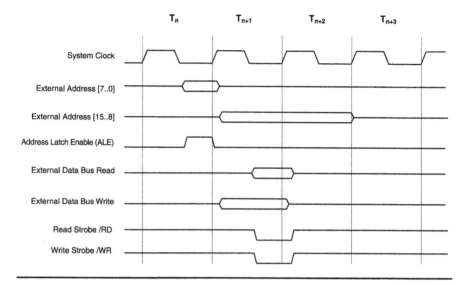

Figure 2-7
External data SRAM memory cycle without wait state.

2.3.3 General-Purpose Register File

Figure 2-8 shows the structure of the general-purpose registers file in the CPU.

All register operations have direct and single-cycle access to all registers. The only exceptions are the five constant arithmetic and logic instructions SBCI, SUBI, CPI, ANDI, and ORI between a constant and a register, and the LDI instruction for load immediate constant data. These instructions apply only to the registers R16..R31.

As shown in Figure 2-8, each register is also assigned a data memory address, mapping them directly into the first 32 locations of the user SRAM area. Although they are not physically implemented as SRAM locations, this memory organization provides great flexibility in access of the registers, as the X,Y, and Z registers can be set to index any register in the file for CPU model 1. In the CPU model 0, only the Z register is available.

2.3.4 I/O Register

The I/O registers define the working conditions of internal and external hardware. Figure 2-9 lists the implemented I/O registers for all four AVR microcontrollers. If a register is implemented, it is marked by a ✓ in the related line.

GP Register File	Address	
R0	0x00	
R1	0x01	
R2	0x02	
...		
R13	0x0D	
R14	0x0E	
R15	0x0F	
R16	0x10	
R17	0x11	
...		
R26	0x1A	X register low byte
R27	0x1B	X register high byte
R28	0x1C	Y register low byte
R29	0x1D	Y register high byte
R30	0x1E	Z register low byte
R31	0x1F	Z register high byte

Figure 2-8
General-purpose register file.

The different I/O registers are accessed by the IN and OUT instructions transferring data between the 32 general-purpose working registers and the I/O space.

I/O registers within the address range 0x00–0x1F are bit-accessible using the SBI and CBI instructions directly. In these registers, the value of single bits can be checked by using the SBIS and SBIC instructions.

2.3.5 EEPROM Data Memory

The microcontrollers of the AVR family contain different sizes of EEPROM memory. This memory is organized as a separate data space, in which single bytes can be read and written. The EEPROM has an endurance of at least 100,000 write/erase cycles.

The EEPROM access registers are accessible in the I/O space using the IN and OUT instructions specifying the EEPROM address register, the EEPROM data register, and the EEPROM control register.

The EEPROM address register specifies the EEPROM address in the related EEPROM space. In Figure 2-10, the register EEARH is implemented only in the AT90S4414 and AT90S8515. The EEPROM data bytes are addressed linearly between 0 and the end of EEPROM.

For the EEPROM write operation, the EEDR register contains the data to be written to the EEPROM in the address given by the EEAR register. For the

Address	Name	Function	1200	2313	4414	8515
0x3F (0x5F)	SREG	Status REGister	✓	✓	✓	✓
0x3E (0x5E)	SPH	Stack Pointer High			✓	✓
0x3D (0x5D)	SPL	Stack Pointer Low		✓	✓	✓
0x3C (0x5C)						
0x3B (0x5B)	GIMSK	General Interrupt MaSK register	✓	✓	✓	✓
0x3A (0x5A)						
0x39 (0x59)	TIMSK	Timer/counter Interrupt MaSK register	✓	✓	✓	✓
0x38 (0x58)	TIFR	Timer/counter Interrupt Flag Register	✓	✓	✓	✓
0x37 (0x57)						
0x36 (0x56)						
0x35 (0x55)	MCUCR	MCU Control Register	✓	✓	✓	✓
0x34 (0x54)						
0x33 (0x53)	TCCR0	Timer/Counter0 Control Register	✓	✓	✓	✓
0x32 (0x52)	TCNT0	Timer/CouNTer0 (8-bit)	✓	✓	✓	✓
0x31 (0x51)						
0x30 (0x50)						
0x2F (0x4F)	TCCR1A	Timer/Counter1 Control Register A		✓	✓	✓
0x2E (0x4E)	TCCR1B	Timer/Counter1 Control Register B		✓	✓	✓
0x2D (0x4D)	TCNT1H	Timer/CouNTer1 High		✓	✓	✓
0x2C (0x4C)	TCNT1L	Timer/CouNTer1 Low		✓	✓	✓
0x2B (0x4B)	OCR1AH	t/c1 Output Compare Register A High		✓	✓	✓
0x2A (0x4A)	OCR1AL	t/c1 Output Compare Register A Low		✓	✓	✓
0x29 (0x49)	OCR1BH	t/c1 Output Compare Register B High			✓	✓
0x28 (0x48)	OCR1BL	t/c1 Output Compare Register B Low			✓	✓
0x27 (0x47)						
0x26 (0x46)						
0x25 (0x45)	ICR1H	t/c1 Input Capture Register High		✓	✓	✓
0x24 (0x44)	ICR1L	t/c1 Input Capture Register Low		✓	✓	✓
0x21 (0x41)	WDTCR	WatchDog Timer Control Register	✓	✓	✓	✓
0x1F (0x3F)	EEARH	EEPROM Address Register High			✓	✓
0x1E (0x3E)	EEARL	EEPROM Address Register Low	✓	✓	✓	✓
0x1D (0x3D)	EEDR	EEPROM Data Register	✓	✓	✓	✓
0x1C (0x3C)	EECR	EEPROM Control Register	✓	✓	✓	✓
0x1B (0x3B)	PORTA	Data Register Port A			✓	✓
0x1A (0x3A)	DDRA	Data Direction Register Port A			✓	✓
0x19 (0x39)	PINA	Input Pins Port A			✓	✓
0x18 (0x38)	PORTB	Data Register Port B	✓	✓	✓	✓
0x17 (0x37)	DDRB	Data Direction Register Port B	✓	✓	✓	✓
0x16 (0x36)	PINB	Input Pins Port B	✓	✓	✓	✓
0x15 (0x35)	PORTC	Data Register Port C			✓	✓

Figure 2-9
I/O register definition.

Address	Name	Function	1200	2313	4414	8515
0x14 (0x34)	DDRC	Data Direction Register Port C			✓	✓
0x13 (0x33)	PINC	Input Pins Port C			✓	✓
0x12 (0x32)	PORTD	Data Register Port D	✓	✓	✓	✓
0x11 (0x31)	DDRD	Data Direction Register Port D	✓	✓	✓	✓
0x10 (0x30)	PIND	Input Pins Port D	✓	✓	✓	✓
0x0F (0x2F)	SPDR	SPI Data Register			✓	✓
0x0E (0x2E)	SPSR	SPI Status Register			✓	✓
0x0D (0x2D)	SPCR	SPI Control Register			✓	✓
0x0C (0x2C)	UDR	UART Data Register		✓	✓	✓
0x0B (0x2B)	USR	UART Status Register		✓	✓	✓
0x0A (0x2A)	UCR	UART Control Register		✓	✓	✓
0x09 (0x29)	UBRR	UART BaudRate Register		✓	✓	✓
0x08 (0x28)	ACSR	Analog Comp. Control & Status Register	✓	✓	✓	✓
0x07 (0x27)						
0x06 (0x26)						
0x05 (0x25)						
0x04 (0x24)						
0x03 (0x23)						
0x02 (0x22)						
0x01 (0x21)						
0x00 (0x20)						

Figure 2-9
Continued.

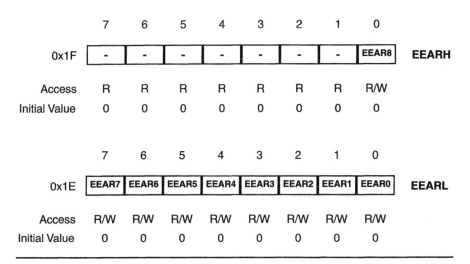

Figure 2-10
EEPROM address register.

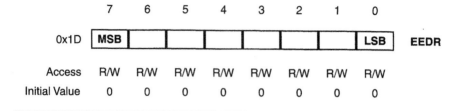

	7	6	5	4	3	2	1	0	
0x1D	MSB							LSB	EEDR
Access	R/W	R/W	R/W	R/W	R/W	R/W	R/W	R/W	
Initial Value	0	0	0	0	0	0	0	0	

Figure 2-11
EEPROM data register.

EEPROM read operation, the EEDR register contains the data read out from the EEPROM at the address given by EEAR. Figure 2-11 shows the details of the EEDR register.

Two bits in the EEPROM control register are responsible for data handling. Figure 2-12 shows the details of the EECR register.

The EEPROM write enable signal (EEWE) is the write strobe to the *EEP-ROM*. When address and data are correctly set up, the EEWE bit must be set to write the value into the EEPROM. When the write access time (typically 2.5 ms at $V_{cc} = 5$ V or 4 ms at $V_{cc} = 2.7$ V) has elapsed, the EEWE bit is cleared (zero) by hardware. The user software can poll this bit and wait for a zero before writing the next byte.

The EEPROM read enable signal (EERE) is the read strobe to the *EEP-ROM*. When the correct address is set up in the EEAR register, the EERE bit must be set. When the EERE bit is cleared (zero) by hardware, requested data is found in the EEDR register. The EEPROM read access time is within a single clock cycle, and there is no need to poll the EERE bit.

The following program example demonstrates writing to an EEPROM.

Depending on the supply voltage, the write access time of the internal EEPROM is in the range of 2.5 to 4 ms. A self-timing function, however, lets the user software detect when the next byte can be written. When the write ac-

	7	6	5	4	3	2	1	0	
0x33	-	-	-	-	-	-	EEWE	EERE	EECR
Access	R	R	R	R	R	R	R/W	R/W	
Initial Value	0	0	0	0	0	0	0	0	

Figure 2-12
EEPROM control register.

AVR RISC Microcontroller Handbook

cess time has elapsed, the EEWE bit is cleared by hardware. In our software example, this bit is polled to wait for a zero before writing the next byte or doing anything else.

The data byte $5A should be stored on address $20 in EEPROM. After the EEAR and EEDR registers are loaded, the EEWE bit in the EECR register must be strobed to start the EEPROM write process. After that, the loop polls the EEWE bit until it is cleared.

```
       ldi   temp, $20    ; Set EEPROM address $20
       out   EEAR, temp
       ldi   temp, $5A    ; Set EEPROM data $5A
       out   EEDR, temp
       ldi   temp, $02    ; Set EEPROM Write Enable
       out   EECR, temp
loop:  in    temp, EECR   ; Read EEPROM Control Register
       sbrc  temp, 1      ; Skip if EEWE bit is cleared
       rjmp  loop         ; Wait until EEWE is cleared
       nop                ; Further instructions out of the loop
```

2.4 Peripherals

2.4.1 Timer/Counter

All of the AVR microcontrollers provide timer/counters. Microcontrollers with CPU model 0 contain only one 8-bit timer/counter; those with CPU model 1 provide one additional 16-bit timer/counter. Both timer/counters have a separate 10-bit prescaler and can be used either as a timer with an internal clock timebase or as a counter with an external pin for counting.

Timer/Counter Prescaler   The 10-bit timer/counter prescaler permits the selection of four different prescaled clocks for both timer/counters. Figure 2-13 shows all details of the timer/counter prescaler. For microcontrollers containing only one timer/counter, the grayed part of the timer/counter prescaler does not exist.

Selections of CK/8, CK/64, CK/256, and CK/1024 are possible where CK is the oscillator clock. For the two timer/counters, additional choices (CK, external source, and stop) can be selected as clock sources.

8-Bit Timer/Counter0   Because of the internal connection to the timer/ counter prescaler, the 8-bit Timer/Counter0 can select its clock source from

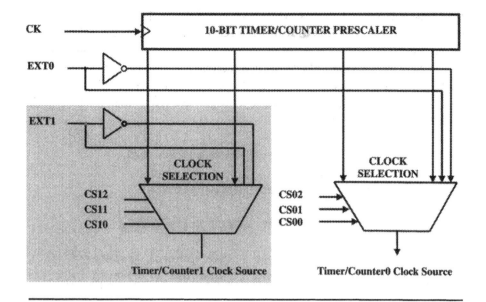

Figure 2-13
Timer/counter prescaler.

CK, prescaled CK, or an external pin. The Timer/Counter0 is realized as an up-counter. Figure 2-14 shows a block diagram of the timer/counter0.

Operation of the Timer/Counter0 is defined by the the Timer/Counter0 control register (TCCR0) shown in Figure 2-15.

The Clock Select0 bits CS02 to CS00 define the prescaled clock source for Timer/Counter0 as described in Table 2-1.

Figure 2-14
Timer/counter0 block diagram.

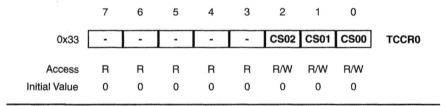

Figure 2-15
Timer/counter0 control register (TCCR0).

Table 2-1
Timer/counter0 prescale select.

CS02	CS01	CS00	Description
0	0	0	Stop Timer/Counter0
0	0	1	CK
0	1	0	CK/8
0	1	1	CK/64
1	0	0	CK/256
1	0	1	CK/1024
1	1	0	External clock at pin T0, rising edge
1	1	1	External clock at pin T0, falling edge

Timer/Counter overflow is signalized by setting the Timer/Counter0 Overflow Flag TOV0, in the timer/counter interrupt flag register (TIFR). Figure 2-16 shows the bits of the Timer/Counter Interrupt Flag register.

The interrupt enable/disable settings for Timer/Counter0 are found in the Timer/Counter Interrupt Mask register (TIMSK) shown in Figure 2-17.

For Timer/Counter0, only the Timer/Counter0 overflow interrupt enable bit is of interest. If this bit is set and the global interrupt (I-bit in the status register) is also set, the Timer/Counter0 overflow interrupt is enabled.

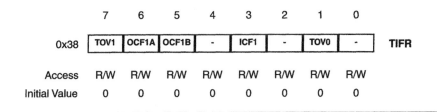

Figure 2-16
Timer/counter interrupt flag register (TIFR).

Figure 2-17
Timer/counter interrupt mask register (TIMSK).

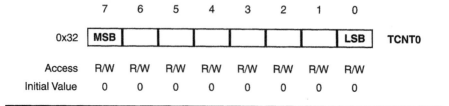

Figure 2-18
Timer/counter0 register (TCNT0).

The Timer/Counter0 is realized as an up-counter with read and write access to the Timer/Counter0 register (TCNT0) (see Figure 2-18).

If the Timer/Counter0 is written and a clock source is present, the Timer/Counter0 continues counting in the clock cycle following the write operation.

When the Timer/Counter0 is externally clocked, the external signal is synchronized with the oscillator frequency of the CPU. To ensure proper sampling of the external clock, the minimum time for the external clock being low and high must be at least one internal CPU clock period. The external clock signal is sampled on the rising edge of the internal CPU clock.

The 8-bit Timer/Counter0 features both high resolution and high accuracy with the lower prescaling opportunities. Similarly, the high prescaling opportunities make the Timer/Counter0 useful for lower speed functions or exact timing functions with infrequent actions.

16-Bit Timer/Counter1   The 16-bit Timer/Counter1 is more complex and is contained only in microcontrollers with CPU model 1. Figure 2-19 shows the block diagram of Timer/Counter1.

The Timer/Counter1 supports an Output Compare function using the Output Compare register1 (OCR1xx) as the data source to be compared to the Timer/Counter1 contents. The Output Compare functions include optional clearing of the counter on compareA matches, and actions on the Output Compare pin 1 on compare matches.

Timer/Counter1 can also be used as a 8-bit, 9-bit, or 10-bit pulse-width modulator. In this mode the counter and the OCR1A/OCR1B registers serve as a glitch-free stand-alone PWM with centered pulses.

The Input Capture function of Timer/Counter1 provides a capture of the Timer/Counter1 contents to the Input Capture register ICR1, triggered by an external event on the Input Capture pin ICP. A noise canceler function can be used to avoid triggering by short spikes.

If the noise canceler function is enabled, the actual trigger condition for the capture event is monitored over four samples before the capture is activated.

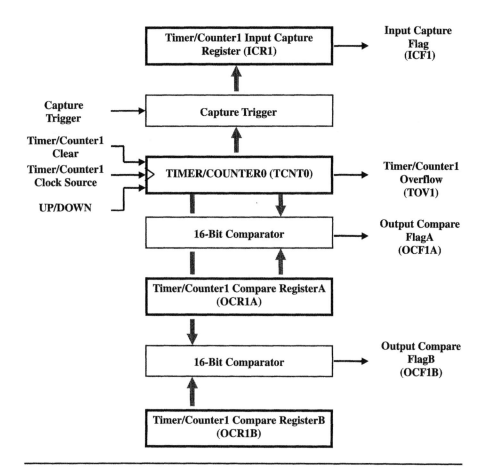

Figure 2-19
Timer/Counter1 block diagram.

The sampling clock is the same clock as the clock source selected for the Timer/Counter1.

The actual capture event settings are defined by the Timer/Counter1 control register (TCCR1B).

Because of the internal connection to the timer/counter prescaler, the 16-bit Timer/Counter1 can select its clock source from CK, prescaled CK, or an external pin. The Timer/Counter1 is normally realized as an up-counter. In the PWM mode it will work as an up- and down-counter.

The Timer/Counter1 control registers TCCR1A and TCCR1B define the operations mode of the Timer/Counter1. Figures 2-20 and 2-21 show both these Timer/Counter1 control registers.

The operation of the Timer/Counter1 is defined by the TCCR1A entries. The bits COM1xx define the output pin action of the Output Compare pins

	7	6	5	4	3	2	1	0	
0x2F	COM1A1	COM1A0	COM1B1	COM1B0	-	-	PWM11	PWM10	**TCCR1A**
Access	R/W	R/W	R/W	R/W	R	R	R/W	R/W	
Initial Value	0	0	0	0	0	0	0	0	

Figure 2-20
Timer/Counter1 control register A (TCCR1A).

	7	6	5	4	3	2	1	0	
0x2E	INCN1	ICES1	-	-	CTC1	CS12	CS11	CS10	**TCCR1B**
Access	R/W	R/W	R	R	R/W	R/W	R/W	R/W	
Initial Value	0	0	0	0	0	0	0	0	

Figure 2-21
Timer/Counter1 control register B (TCCR1B).

following a compare match as Table 2-2 defines. Because the Output Compare pin is an alternative function of an I/O port, the Data Direction bit must be set for output.

The PWM functionality is defined by the PWM1x bits as shown in Table 2-3. The PWM operation is described later in this chapter.

The Timer/Counter1 control register B controls some further features of the Timer/Counter1.

The Input Capture Noise Canceler samples four successive measures on the Input Capture pin. All samples must be high or low relative to the selected trigger edge. To enabled the input Capture Noise Canceler function, the INCN1

Table 2-2
Output Compare1 mode select.

COM1x1	COM1x0	Description
0	0	Disconnect Timer/Counter1 from pin OC1x
0	1	Toggle OC1x line
1	0	Clear the OC1x line
1	1	Set the OC1x line

Table 2-3
PWM mode select.

PWM11	PWM10	Description
0	0	Disable PWM operation of Timer/Counter1
0	1	Timer/Counter1 is an 8-bit PWM
1	0	Timer/Counter1 is an 9-bit PWM
1	1	Timer/Counter1 is an 10-bit PWM

bit must be set. If this bit is cleared, the Input Capture Noise Canceler function is disabled, and for triggering one edge is sufficient.

The edge of the trigger signal on the Input Capture pin is defined by the Input Capture Edge Select bit ICES1. If this bit is set, the rising edge of an input capture signal will trigger.

When the CTC1 control bit is set, the Timer/Counter1 is reset to 0x0000 in the clock cycle after a compareA match.

If the CTC1 control bit is cleared, the Timer/Counter1 continues counting until it is stopped, cleared, wraps around (overflow), or changes direction. In PWM mode, this bit has no effect.

The Clock Select1 bits CS12 to CS10 define the prescaled clock source for Timer/Counter1 as described in Table 2-4.

Different status flags (Overflow, Compare Match, and Capture Event) for Timer/Counter1 are arranged in the Timer/Counter Interrupt Flag register (TIFR). Figure 2-22 shows the bits of this register.

Table 2-4
Timer/Counter1 prescale select.

CS12	CS11	CS10	Description
0	0	0	Stop Timer/Counter
0	0	1	CK
0	1	0	CK/8
0	1	1	CK/64
1	0	0	CK/256
1	0	1	CK/1024
1	1	0	External Clock at Pin T1, rising edge
1	1	1	External Clock at Pin T1, falling edge

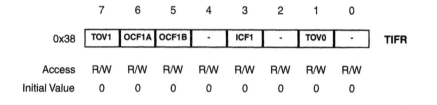

	7	6	5	4	3	2	1	0	
0x38	TOV1	OCF1A	OCF1B	-	ICF1	-	TOV0	-	**TIFR**
Access	R/W	R/W	R/W	R/W	R/W	R/W	R/W	R/W	
Initial Value	0	0	0	0	0	0	0	0	

Figure 2-22
Timer/Counter interrupt flag register (TIFR).

The interrupt enable/disable setting for Timer/Counter1 are found in the Timer/Counter Interrupt Mask register (TIMSK) shown in Figure 2-23.

For Timer/Counter1, there are four interrupt flags and four interrupt enable bits. If an interrupt enable bit is set, the regarding interrupt is enabled only if the global interrupt bit (I-bit in the status register) is also set.

The Timer/Counter1 Overflow Flag TOV1 is set when an overflow occurs in Timer/Counter1. Alternatively, TOV1 is cleared by writing a logic 1 to the flag. In PWM mode, this bit is set when Timer/Counter1 changes counting direction at 0x0000.

The Output Compare flag 1 (OCF1A/OCF1B) bit is set when compare match occurs between the Timer/Counter1 and the data in output compare register 1 (OCR1A/OCR1B).

The Input Capture flag 1 (ICF1) bit is set to flag an input capture event, indicating that the Timer/Counter1 value has been transferred to the input capture register ICR1.

The Interrupt flags are cleared by hardware when executing the corresponding interrupt handling vector. Alternatively, these flags are cleared by writing a logic 1 to the flag.

The Timer/Counter1 is realized as a counter with read and write access to the Timer/Counter1 register TCNT1 (see Figure 2-24).

	7	6	5	4	3	2	1	0	
0x39	TOIE1	OCIE1	OCIE1B	-	TICIE1	-	TOIE0	-	**TIMSK**
Access	R/W	R/W	R/W	R	R/W	R	R/W	R	
Initial Value	0	0	0	0	0	0	0	0	

Figure 2-23
Timer/Counter Interrupt Mask register (TIMSK).

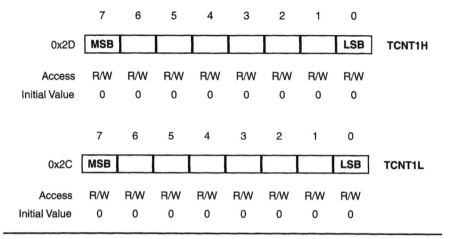

Figure 2-24
Timer/Counter1 register TCNT1H/TCNT1L.

If the Timer/Counter1 is written and a clock source is present, the Timer/Counter1 continues counting in the clock cycle following the write operation.

When Timer/Counter1 is externally clocked, the external signal is synchronized with the oscillator frequency of the CPU. To ensure proper sampling of the external clock, the minimum time for the external clock being low and high must be at least one internal CPU clock period. The external clock signal is sampled on the rising edge of the internal CPU clock.

The 16-bit Timer/Counter1 features both high resolution and high accuracy with the lower prescaling opportunities. Similarly, the high prescaling opportunities make the Timer/Counter1 useful for lower speed functions or exact timing functions with infrequent actions.

Timer/Counter1 in PWM Mode   When the PWM mode is selected, Timer/Counter1, the Output Compare register1A (OCR1A), and the Output Compare register1B (OCR1B), form a dual 8, 9, or 10-bit, free-running, glitch-free and phase-correct pulse-width modulator with outputs on the OC1A and OC1B pins. Timer/Counter1 acts as an up/down counter, counting up from zero to the top value, then turning and counting down again to zero before the cycle is repeated. Table 2-5 shows the selection of resolution and the resulting frequency of the pulse-width-modulated period. The frequency $f_{T/C1}$ means the timer clock frequency (after prescaler).

When the counter value matches the contents of the 10 least significant bits of OCR1A or OCR1B, the OC1A/OC1B pins are set or cleared according to the settings of the COM1X1/COM1X0 bits in the TCCR1A register. Table 2-6

Table 2-5

Timer/Counter1 PWM parameters.

PWM Resolution	Timer Top Value	Frequency
8-bit	0x00FF	$f_{T/C1}/510$
9-bit	0x01FF	$f_{T/C1}/1022$
10-bit	0x03FF	$f_{T/C1}/2046$

Table 2-6

Compare1 mode select in PWM mode.

COM1x1	COM1x0	Effect on OCX1
0	0	Not connected.
0	1	Not connected.
1	0	Cleared on compare match during up-counting. Set on compare match during down-counting.
1	1	Set on compare match during down-counting. Cleared on compare match during up-counting.

shows the effects on PWM output in relation to the COM1X1/COM1X0 settings.

To avoid odd-length PWM pulses (glitches), the 10 least significant OCR1A/OCR1B bits are transferred to a temporary location after any change. They are latched when Timer/Counter1 reaches the top value. Figure 2-25 shows the difference in time between changing and activating the compare values and the phases of the output signal relating to Table 2-6.

In PWM mode, the Timer Overflow flag TOV1 is set when the counter changes direction at $0000. Timer Overflow Interrupt1 operates exactly as in normal Timer/Counter mode, that is, it is executed when TOV1 is set, provided that Timer Overflow Interrupt1 and global interrupts are enabled. This also applies to the Timer Output Compare1 flags and interrupts.

The PWM output frequency is given by

$$f_{PWM} = \frac{f_{TCI}}{2046},$$

where f_{TCI} is the Timer/Counter1 clock source frequency.

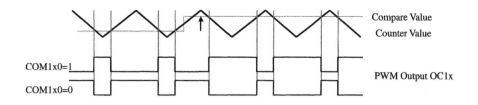

Figure 2-25
PWM pulse generation.

2.4.2 Watchdog Timer

The watchdog timer checks the error-free operation of the microcontroller. In the case of a correctly running program, the watchdog timer must be reset by the program with the instruction wdr before it will overflow. Otherwise, the watchdog timer resets the microcontroller and executes from the reset vector.

The watchdog timer is clocked from a separate on-chip oscillator that runs at 1 MHz. The watchdog reset interval can be adjusted from 16 to 2048 ms by controlling the watchdog timer prescaler with the register WDTCR.

Figure 2-26 shows a block diagram of the watchdog timer.

From the watchdog reset, eight different clock cycle periods can be selected to determine the reset period.

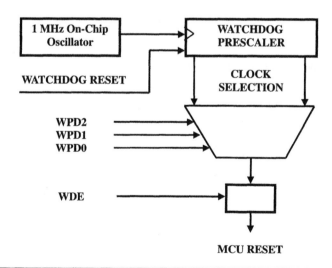

Figure 2-26
Watchdog timer.

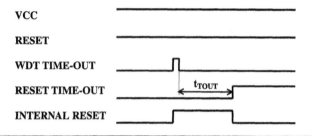

Figure 2-27
Watchdog reset timing.

The timing for a watchdog timer reset is shown in Figure 2-27. If the reset period expires without another watchdog reset, the watchdog will generate a short reset pulse of one XTAL cycle duration. On the falling edge of this pulse, the delay timer starts counting the time-out period t_{TOUT}, typically 16 ms.

The different conditions for the watchdog timer are defined by the contents of the watchdog timer control register WDTCR. Figure 2-28 shows the features of the register WDTCR.

The Watchdog Enable bit WDE enables the watchdog function if set. Otherwise, the watchdog is disabled and no check of program flow occurs.

The bits WDP2..0 determine the watchdog timer prescaling when the watchdog timer is enabled. The different prescaling values and their corresponding timeout periods are shown in Table 2-7.

2.4.3 Serial Peripheral Interface

The serial peripheral interface (SPI) allows high-speed synchronous data transfer between an AVR microcontroller and peripheral devices or between several AVR microcontrollers. An SPI device can operate as master or as slave.

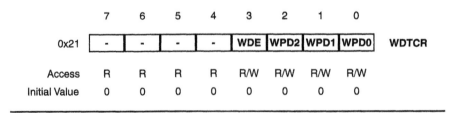

Figure 2-28
Watchdog timer control register (WDTCR).

Table 2-7
Watchdog timer prescale select.

WDP2	WDP1	WDP0	Timeout Period (ms)
0	0	0	16
0	0	1	32
0	1	0	64
0	1	1	128
1	0	0	256
1	0	1	512
1	1	0	1024
1	1	1	2048

The master generates the clock for synchronous data transmission. Therefore, in master mode SCK operates as clock output and in slave mode as clock input.

The implemented SPI features include the following:

• Full-duplex, three-wire synchronous data transfer
• Master or slave operation
• 5 Mbit/s bit frequency maximum
• LSB first or MSB first data transfer
• Four programmable bit rates
• End of transmission interrupt request
• Write collision protection
• Wakeup from idle mode (slave mode only)

To accomplish communication, typically four pins are used:

• MOSI—Master-out slave-in
• MISO—Master-in slave-out
• SCK—Serial clock
• /SS—Slave select

Figure 2-29 shows the connection of two devices via the SPI.

Writing to the SPI Data register (SPDR) of the master device starts the SPI clock generator, and the data written shifts out of the MOSI pin and into the MOSI pin of the slave device. After shifting 1 byte, the SPI clock generator stops, setting the End-of-Transmission flag (SPIF) in the SPI Status register

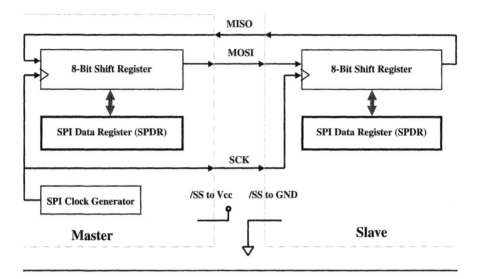

Figure 2-29
SPI master–slave connection.

(SPSR). If the SPI Interrupt Enable bit (SPIE) in the SPCR register is set, an interrupt is requested.

The Slave Select input /SS is set low to select an individual SPI device as a slave. When /SS is set high, the SPI port is deactivated and the MOSI pin can be used as an input. Slave/master mode can also be selected in software by clearing or setting the MSTR bit in the SPI Control register (SPCR).

The two shift registers in the master and the slave can be considered as one distributed 16-bit circular shift register. When data is shifted from the master to the slave, data is also shifted in the opposite direction simultaneously. This means that during one shift cycle, data in the master and the slave is interchanged.

The control of synchronous data transmission is organized entirely by three SPI registers. The SPI Control register contains some bits to control the data transmission. Figure 2-30 shows the details.

The SPI Interrupt Enable bit (SPIE) enables SPI interrupt requests provided that global interrupts are enabled. Setting the SPI Enable bit (SPE) causes an internal connection of SS, MOSI, MISO, and SCK to the pins PB4, PB5, PB6, and PB7.

The setting of the Data Order bit (DORD) defines the order of bit transmitted. When DORD is set, the LSB of the data word is transmitted first. When DORD is cleared, the MSB of the data word is transmitted first.

AVR RISC Microcontroller Handbook

0x0D	7	6	5	4	3	2	1	0	
	SPIE	SPE	DORD	MSTR	CPOL	CPHA	SPR1	SPR0	SPCR
Access	R/W	R/W	R/W	R/W	R/W	R/W	R/W	R/W	
Initial Value	0	0	0	0	0	0	0	0	

Figure 2-30
SPI Control register (SPCR).

The Master/Slave Select bit (MSTR) selects master SPI mode when set, and slave SPI mode when cleared. If the SS pin is set low by another device (or hardware) while MSTR is set, then MSTR will be forced low and the device will be a slave.

Clock polarity and clock phase can be defined very flexibly. Figures 2-31 and 2-32 explain the settings in detail.

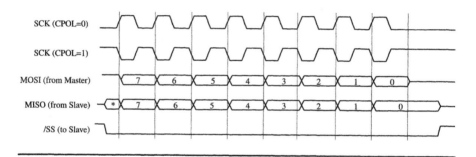

Figure 2-31
SPI transfer format with CPHA=0.

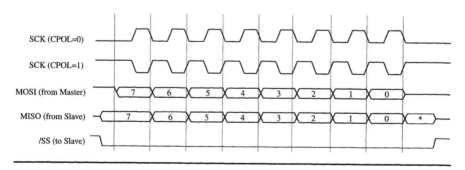

Figure 2-32
SPI transfer format with CPHA=1.

Table 2-8
SCK clock selection.

SPR1	SPR0	Clock
0	0	CK/4
0	1	CK/16
1	0	CK/64
1	1	CK/128

The SPI Clock Rate Select bits (SPRx) control the SCK rate of the device configured as master. SPR1 and SPR2 have no effect on the slave. The relationship between SCK and the oscillator clock frequency CK is shown in Table 2-8.

The SPI Status register SPSR contains two relevant bits only. Figure 2-33 shows the definition of this register.

After completion of a serial transfer, the SPIF bit is set and an interrupt is requested if SPI and global interrupts are enabled. SPIF is cleared by hardware when executing the corresponding interrupt handling vector. Alternatively, the SPIF bit is cleared by first reading the SPI Status register with SPIF set, then accessing the SPI Data register (SPDR).

The Write Collision flag (WCOL) is set if the SPDR is written during a data transfer. During data transfer, the result of reading the SPDR may be incorrect, and writing to it will have no effect.

WCOL is cleared by first reading the SPI Status register with WCOL set, and then accessing the SPI Data register.

The SPI Data register shown in Figure 2-34 is a read/write register used for data transfer between the register file and the SPI Shift register. Writing to the register initiates data transmission. Reading the register causes the Shift Register Receive buffer to be read.

	7	6	5	4	3	2	1	0	
0x0E	SPIF	WCOL	-	-	-	-	-	-	SPSR
Access	R	R	R	R	R	R	R	R	
Initial Value	0	0	0	0	0	0	0	0	

Figure 2-33
SPI status register (SPSR).

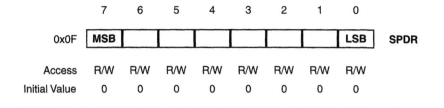

Figure 2-34
SPI data register (SPDR).

The system is single-buffered in the transmit direction and double-buffered in the receive direction internally. This means that characters to be transmitted cannot be written to the SPI data register before the entire shift cycle is completed. When data is being received, however, a received character must be read from the SPI data register before the next character has been completely shifted in. Otherwise, the first character is lost.

When the SPI is enabled, the DDB5-DDB7 bits in data direction register B (DDRB) are overridden and have no effect. The SPI interface is also used for flash memory and EEPROM downloading and uploading, respectively.

2.4.4 Universal Asynchronous Receiver and Transmitter

Devices of the AVR microcontroller family containing CPU model 1 feature a full-duplex universal asynchronous receiver and transmitter (UART).

The main features of the UART are as follows:

• Baud rate generator generates any baud rate
• High baud rates at low oscillator frequencies
• 8 or 9 bits data
• Noise filtering by oversampling
• Overrun detection
• Framing error detection
• False start bit detection
• Three separate interrupts on TX Complete, TX Data Register Empty, and RX Complete

Data Transmission   Figure 2-35 shows a simplified block schematic of the transmitting section.

Data transmission is initiated by writing the data to be transmitted to the UART I/O Data register (UDR). Data is transferred from the UDR to the Transmit Shift register when the stop bit of the character currently being

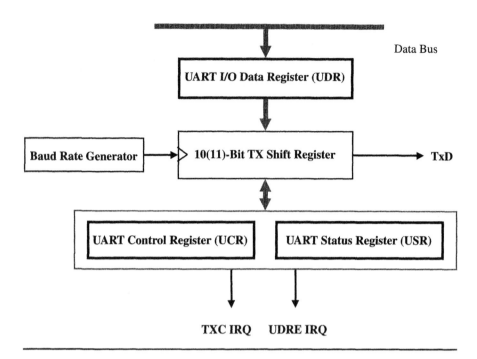

Figure 2-35
UART transmitter (simplified).

transmitted has been shifted out. If the Transmit Shift register is empty, it will be loaded immediately.

After the data is transferred from the UDR to the Transmit Shift register, the UDRE (UART Data Register Empty) bit in the UART Status register (USR) is set. When this bit is set, the UART is ready to receive the next character.

At the same time as the data is transferred from UDR to the 10(11)-bit shift register, bit 0 of the shift register is cleared (start bit) and bit 9 or 10 is set (stop bit). If a 9-bit data word is selected (the CHR9 bit in the UART control register, UCR, is set), the TXB8 bit in UCR is transferred to bit 9 in the Transmit Shift register.

On the baud rate clock following the transfer operation to the shift register, the start bit is shifted out on the TXD pin. Then follow the data bits follow, with LSB first.

When the stop bit has been shifted out, the shift register is loaded if any new data has been written to the UDR during the transmission.

During loading, UDRE is set. If there is no new data in the UDR register to send when the stop bit is shifted out, the UDRE flag will remain set. In this case, after the stop bit has been present on TXD for one bit length, the TX Complete flag (TXC) in the USR register is set.

AVR RISC Microcontroller Handbook

The TXEN bit in the UCR register enables the UART transmitter when set. When this bit is cleared, the pin PD1 can be used for general I/O. When TXEN is set, the UART transmitter will be connected to pin PD1, regardless of the setting of the DDD1 bit in DDRD.

Data Reception Figure 2-36 shows a simplified block schematic of the reception section.

The receiver front-end logic (not represented in Figure 2-36) samples the signal on the RXD pin at a frequency 16 times the baud rate. Figure 2-37 shows the details of the sampling process.

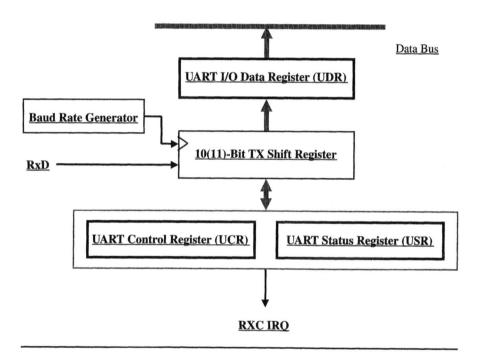

Figure 2-36
UART receiver (simplified).

Figure 2-37
Sampling received data.

While the line is idle, one single sample of logical zero will be interpreted as the falling edge of a start bit, and the start bit detection sequence is initiated. Following this 1–0 transition, the receiver samples the RxD pin at samples 8, 9, and 10.

If, two or more of these three samples are found to be logical ones, the start bit is rejected as a noise spike and the receiver starts looking for the next falling edge.

If, however, a valid start bit is detected, sampling of the data bits following this start bit is performed. These bits are also sampled at samples 8, 9, and 10. The logical value found in at least two of the three samples is taken as the bit value. All bits are shifted into the transmitter shift register as they are sampled.

When the stop bit enters the receiver, at least two of the three samples must be 1 to accept the stop bit. If two or more samples are logical zeros, the Framing Error (FE) flag in the UART Status register is set. Before reading the UDR register, the user should always check the FE bit to detect framing errors. Whether or not a valid stop bit is detected at the end of a character reception cycle, the data is transferred to the UDR and the RXC flag in USR is set.

The UDR is, in fact two physically separate registers, one for transmitted data and one for received data. When the UDR is read, the Receive Data register is accessed, and when the UDR is written, the Transmit Data register is accessed.

If a 9-bit data word is selected (the CHR9 bit in the UART Control register is set), the RXB8 bit in the UCR is loaded with bit 9 in the Transmit shift register when data is transferred to UDR.

If, after having received a character, the UDR register has not been accessed since the last receive, the Overrun flag (OR) in the UCR is set. This means that the new data transferred to the shift register has overwritten the old data not yet read, and the old data is lost. The user should always check the OR bit before reading from the UDR register in order to detect any overruns.

Clearing the RXEN bit in the UCR register, disables the receiver. This means that the pin PD0 can be used as a general I/O pin. When RXEN is set, the UART receiver will be connected to the PD0 pin regardless of the setting of the DDD0 bit in DDRD.

UART Control Figures 2-35 and 2-36 have shown the three registers involved in data transmission and reception of this UART. Definitions and contents of these three registers are shown in Figures 2-38 to 2-40.

The UART Data register is the physical interface between the databus and the transmission and reception shift registers. It is implemented twice under the same I/O address. When writing to the UDR, the UART Transmit Data register is written. When reading from UDR, the UART Receive Data register is read.

AVR RISC Microcontroller Handbook

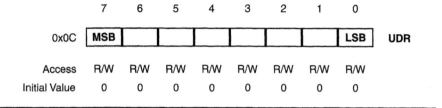

Figure 2-38
UART I/O data register (UDR).

0x0C	7	6	5	4	3	2	1	0	UDR
	MSB							LSB	
Access	R/W	R/W	R/W	R/W	R/W	R/W	R/W	R/W	
Initial Value	0	0	0	0	0	0	0	0	

Figure 2-39
UART status register (USR).

0x0B	7	6	5	4	3	2	1	0	USR
	RXC	TXC	UDRE	FE	OR	-	-	-	
Access	R	R	R	R	R	R	R	R	
Initial Value	0	1	1	0	0	0	0	0	

Figure 2-40
UART control register (UCR).

0x0A	7	6	5	4	3	2	1	0	UCR
	RXCIE	TXCIE	UDRIE	RXEN	TXEN	CHR9	RXB9	TXB9	
Access	R/W	R/W	R/W	R/W	R/W	R/W	R	R/W	
Initial Value	0	0	0	0	0	0	0	0	

The bits in the UART Status register represent results of data interchange. Therefore, they are only readable.

The Receive Complete bit (RXC) is set when a received character is transfered from the Receive Shift register to UDR.

When the RXCIE bit in the UCR is set, setting of RXC causes the UART Receive Complete interrupt to be executed. RXC is cleared by hardware when executing the corresponding interrupt handling vector. Alternatively, the bit is cleared by first reading the USR while RXC is set and then reading UDR.

The Transmit Complete bit is set when the entire character (including the stop bit) in the Transmit Shift register has been shifted out and no new data has been written to the UDR for further transmission. This flag is especially useful in half-duplex communications interfaces, where a transmitting application must enter receive mode and free the communications bus immediately after completing the transmission.

When the TXCIE bit in the UCR is set, setting of TXC causes the UART Transmit Complete interrupt to be executed. TXC is cleared by hardware when executing the corresponding interrupt handling vector. Alternatively, the TXC bit is cleared by first reading the USR while TXC is set and then writing the UDR. This bit is set during reset to indicate that the transmitter is not busy transmitting anything.

The UART Data Register Empty bit (UDRE) is set when a character written to the UDR is transferred to the Transmit shift register. Setting of this bit indicates that the transmitter is ready to receive a new character for transmission.

When the UDRIE bit in the UCR is set, setting of UDRE causes the UART Transmit Complete interrupt to be executed. UDRE is cleared by hardware when executing the corresponding interrupt handling vector. Alternatively, the UDRE bit is cleared by first reading USR while UDRE is set and then writing the UDR. UDRE is set during reset to indicate that the transmitter is ready.

The Framing Error bit (FE) is set when the stop bit of an incoming character is zero (framing error condition). The FE bit is cleared by first reading the USR while FE is set and then reading the UDR.

The Overrun bit (OR) is set when a character already present in the UDR register is not read before the next character is transferred from the Receiver Shift register. The unread character in the UDR will be overwritten. The OR bit is cleared by first reading the USR while OR is set and then reading the UDR.

The bits in the UART control register (UCR) enable different modes of the UART and help to handle the ninth data bit. This ninth data bit can be used as an extra stop bit or a parity bit.

When the RX/TX Complete Interrupt Enable bit (RXCIE or TXCIE) is set, a setting of the RXC/TXC bit in the USR will cause the Receive/Transmit Complete interrupt routine to be executed, provided that global interrupts are enabled. It is possible to enable one or both interrupts.

When the UART Data Register Empty Interrupt Enable bit (UDRIE) is set, a setting of the UDRE bit in the USR will cause the UART Data Register Empty interrupt routine to be executed, provided that global interrupts are enabled.

The Receiver Enable bit (RXEN) enables the UART receiver when set. When the receiver is disabled, the TXC, OR, and FE status flags cannot become set. If these flags are set, turning off RXEN does not cause them to be cleared.

The Transmitter Enable bit (TXEN) enables the UART transmitter when set. When the transmitter is disabled while transmitting a character, the transmitter is not disabled until the character in the shift register plus any following character in the UDR has been completely transmitted.

When the 9-Bit-Characters bit (CHR9) is set, transmitted and received characters are 9 bits long plus start and stop bits. The ninth bit is read and written by using the RXB8 and TXB8 bits in the UCR, respectively. When CHR9 is set, RXB8 is the ninth data bit of the received character and TXB8 is the ninth data bit in the character to be transmitted.

Baud Rate Generator The baud rate generator is a frequency divider that generates the required clock for data transmission and reception.

The resulting baud rate can be calculated according to the following equation:

$$BAUD = \frac{f_{CK}}{16(UBRR + 1)}.$$

UBRR is the contents of the UART Baud Rate register (0 to 255).

For standard crystal frequencies f_{CK}, the most commonly used baud rates can be generated by using the UBRR settings in Table 2-9. UBRR values that yield an actual baud rate differing less than 2% from the target baud rate are given in bold type.

The UBRR register, described in Figure 2-41, is an 8-bit read/write register that specifies the UART baud rate according to the description in Table 2-9.

2.4.5 Analog Comparator

An analog comparator compares two analog input voltages. Its binary output signalizes which of the two voltages is higher.

The AVR microcontrollers contain one analog comparator that compares the input voltages on pin PB2 (AIN0) and pin PB3 (AIN1). A block diagram of the analog comparator is shown in Figure 2-42.

When the voltage on pin AIN0 is higher than the voltage on pin AIN1, the Analog Comparator Output (ACO) is set. The comparator's output can be set to trigger the Timer/Counter1 Input Capture function.

In addition, the comparator can trigger a separate interrupt, exclusive to the analog comparator. The user can select interrupt triggering on comparator output rise, fall, or toggle.

Table 2-9
UBRR settings at various crystal frequencies.

Baud Rate	1MHz	% Error	1.8432MHz	% Error	2 MHz	% Error	2.4576 MHz	% Error
2400	25	0.2	47	0.0	51	0.2	63	0.0
4800	12	0.2	23	0.0	25	0.2	31	0.0
9600	6	7.5	11	0.0	12	0.2	15	0.0
14400	3	7.8	7	0.0	8	3.7	10	3.1
19200	2	7.8	5	0.0	6	7.5	7	0.0
28800	1	7.8	3	0.0	3	7.8	4	6.3
57600	0	7.8	1	0.0	1	7.8	2	12.5
115200	0	84.3	0	0.0	0	7.8	0	25.0

Baud Rate	3.2768 MHz	% Error	3.6864 MHz	% Error	4 MHz	% Error	4.608 MHz	% Error
2400	84	0.4	95	0.0	103	0.2	119	0.0
4800	42	0.8	47	0.0	51	0.2	59	0.0
9600	20	1.6	23	0.0	25	0.2	29	0.0
14400	13	1.6	15	0.0	16	2.1	19	0.0
19200	10	3.1	11	0.0	12	0.2	14	0.0
28800	6	1.6	7	0.0	8	3.7	9	0.0
57600	3	12.5	3	0.0	3	7.8	4	0.0
115200	1	12.5	1	0.0	1	7.8	2	20.0

Baud Rate	7.3728 MHz	% Error	8 MHz	% Error	9.216 MHz	% Error	11.059 MHz	% Error
2400	191	0.0	207	0.2	239	0.0	(287)	—
4800	95	0.0	103	0.2	119	0.0	143	0.0
9600	47	0.0	51	0.2	59	0.0	71	0.0
14400	31	0.0	34	0.8	39	0.0	47	0.0
19200	23	0.0	25	0.2	29	0.0	35	0.0
28800	15	0.0	16	2.1	19	0.0	23	0.0
57600	7	0.0	8	3.7	9	0.0	11	0.0
115200	3	0.0	3	7.8	4	0.0	5	0.0

Baud Rate	14.746 MHz	% Error	16 MHz	% Error	18.432 MHz	% Error	20 MHz	% Error
2400	(383)	—	(416)	—	(479)	—	(520)	—
4800	191	0.0	207	0.2	239	0.0	(259)	—
9600	95	0.0	103	0.2	119	0.0	129	0.2
14400	63	0.0	68	0.6	79	0.0	86	0.2
19200	47	0.0	51	0.2	59	0.0	64	0.2
28800	31	0.0	34	0.8	39	0.0	42	0.9
57600	15	0.0	16	2.1	19	0.0	21	1.4
115200	7	0.0	8	3.7	9	0.0	10	1.4

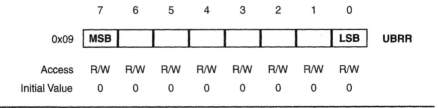

Access R/W R/W R/W R/W R/W R/W R/W R/W
Initial Value 0 0 0 0 0 0 0 0

Figure 2-41
UART baud rate register (UBRR).

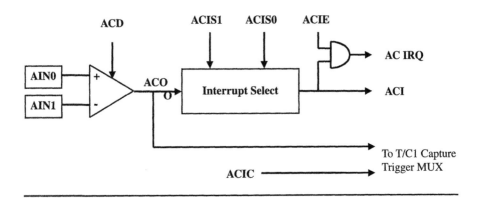

Figure 2-42
Analog comparator block diagram.

The interrupt behavior is controlled by some bits in the Analog Comparator Control and Status register (ACSR). The contents of this register is shown in Figure 2-43.

If the Analog Comparator Disable bit (ACD) is set, the power to the analog comparator is switched off. This bit can be set at any time to turn off the analog comparator. It is most commonly used if power consumption during idle mode is critical and wake-up from the analog comparator is not required. When changing the ACD bit, the Analog Comparator Interrupt must be disabled by clearing the ACIE bit in the ACSR first. Otherwise, an undesirable interrupt can occur when the bit is changed.

The Analog Comparator Output (ACO) is directly connected to the comparator output.

The Analog Comparator Interrupt flag (ACI) is set when a comparator output event triggers the interrupt mode defined by ACIS1 and ACIS0 (see Table 2-10).

AVR RISC Microcontroller Handbook

	7	6	5	4	3	2	1	0	
0x08	ACD	-	ACO	ACI	ACIE	ACIC	ACIS1	ACIS0	ACSR
Access	R/W	R	R	R/W	R/W	R/W	R/W	R/W	
Initial Value	0	0	0	0	0	0	0	0	

Figure 2-43
Analog Comparator Control and Status register (ACSR).

The Analog Comparator Interrupt routine is executed if the ACIE bit and the I-bit in SREG are both set. ACI is cleared by hardware when executing the corresponding interrupt handling vector. Alternatively, ACI is cleared by writing a logic one to the flag.

While the Analog Comparator Input Capture Enable bit (ACIC) is set, the Input Capture function in Timer/Counter1 is triggered by the analog comparator. The comparator output in this case is directly connected to the Input Capture front-end logic, making the comparator utilize the noise canceler and edge select features of the Timer/Counter1 Input Capture interrupt. If ACIC is cleared, no connection between the analog comparator and the Input Capture function is given. To make the comparator trigger the Timer/Counter1 Input Capture interrupt, the TICIE1 bit in the Timer Interrupt Mask register (TIMSK) must be set.

The Analog Comparator Interrupt Mode Select bits (ACIS1 and ACIS0) determine which comparator events trigger the Analog Comparator interrupt. The different settings are shown in Table 2-10.

Table 2-10
ACISx settings.

ACIS1	ACIS0	Interrupt Mode
0	0	Comparator interrupt on output toggle
0	1	Reserved
1	0	Comparator interrupt on falling edge of ACO
1	1	Comparator interrupt on rising edge of ACO

When the ACIS1/ACIS0 bits are changed, the Analog Comparator Inter-rupt must be disabled by clearing its Interrupt Enable bit in the ACSR register. Otherwise, an interrupt can occur when the bits are changed.

2.4.6 I/O Ports

The various devices of the AVR microcontroller family are differently equipped with I/O ports. The Appendix shows the pin diagrams for the AVR microcontrollers described in this book.

At minimum, an AVR microcontroller is equipped with the two digital I/O ports PORTB and PORTD. Some of the pins have an alternative function.

Table 2-11 shows primary and alternative pin functions for the AT90S1200 and AT90S2313 microcontrollers (20-pin devices). Table 2-12 shows these pin functions for the AT90S4414 and AT90S8515 microcontrollers (40-pin devices). The basis for Tables 2-11 and 2-12 are the PDIP packages of those microcontrollers described in the Appendix.

Table 2-11
Pin functions for 20-pin AVR microcontrollers

Pin	Primary Function 20-Pin Device	Alternative Function AT90S1200	Alternative Function AT90S2313
1	/RESET		
2	PD0		
3	PD1		
4	XTAL2		
5	XTAL1		
6	PD2	INT0	INT0
7	PD3		INT1
8	PD4	T0	T0
9	PD5		T1
10	GND		
11	PD6		ICP
12	PB0	AIN0	AIN0
13	PB1	AIN1	AIN1
14	PB2		
15	PB3		OC1
16	PB4		
17	PB5	MOSI	MOSI
18	PB6	MISO	MISO
19	PB7	SCK	SCK
20	VCC		

Table 2-12

Pin functions for 40-pin AVR microcontrollers.

Pin	Primary Function 40-Pin Device	Alternative Function AT90S4414/8515
1	PB0	T0
2	PB1	T1
3	PB2	AIN0
4	PB3	AIN1
5	PB4	/SS
6	PB5	MOSI
7	PB6	MISO
8	PB7	SCK
9	/RESET	
10	PD0	RXD
11	PD1	TXD
12	PD2	INT0
13	PD3	INT1
14	PD4	
15	PD5	OC1A
16	PD6	/WR
17	PD7	/RD
18	XTAL2	
19	XTAL1	
20	GND	
21	PC0	A8
22	PC1	A9
23	PC2	A10
24	PC3	A11
25	PC4	A12
26	PC5	A13
27	PC6	A14
28	PC7	A15
29	OC1B	
30	ALE	
31	ICP	
32	PA7	AD7
33	PA6	AD6
34	PA5	AD5
35	PA4	AD4
36	PA3	AD3
37	PA2	AD2
38	PA1	AD1
39	PA0	AD0
40	VCC	

The alternative pin functions are described in the special chapters considering the related function groups. Therefore, this chapter is reserved for an explanation of the digital I/O function of the ports.

The primary function of each port pin is bidirectional digital I/O and will be controlled by some internal glue logic. Figure 2-44 shows the design principle of the pin circuitry.

Whether a pin operates as output or as input is controlled by the Data Direction bits (DDxn). The character "x" describes the port (A to D) and the character "n" the pin number (0 to 7).

The output is buffered in the PORTxn latch. The characters "x" and "n" describe the same letters and numbers as before.

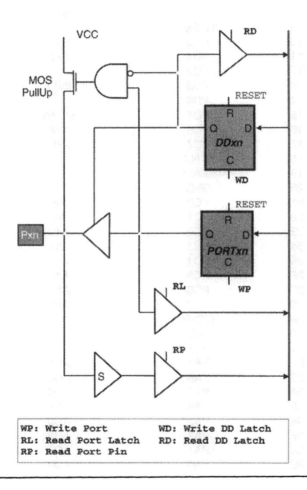

WP: Write Port	WD: Write DD Latch
RL: Read Port Latch	RD: Read DD Latch
RP: Read Port Pin	

Figure 2-44
I/O port principle.

Table 2-13
Pin states for digital I/O.

DDxn	PORTxn	I/O	Pull-up	Comment
0	0	Input	No	Tri-State (Hi-Z)
0	1	Input	Yes	Pin will source current if external pulled low
1	0	Output	No	Push-Pull Zero Output (Lo)
1	1	Output	No	Push-Pull One Output (Hi)

With these two latches, we will get four different pin states for the I/O circuitry. Table 2-13 shows these pin states and the resulting functionality.

If the pin is configured as input, the content of the PORTxn latch defines the operation of the internal Pull-up resistor. Both latches can be written and read back. The I/O pin can also be read directly, without any latch.

All port pins have individually selectable pull-ups. The output buffers can sink 20 mA and, thus, drive LED displays directly. When pins Px0 to Px7 are used as inputs and are externally pulled low, they will source current (IIL) if the internal pull-ups are activated.

The Portx Input Pins address PINx is not a register. This address enables access to the physical value on each Portx pin. When reading PORTx, the Portx Data Latch is read, and when reading PINx, the logical values present on the pins are read. Figure 2-44 shows this difference clearly.

PortA (AT90S4414 and AT90S8515 Only)   Three data memory address locations are allocated for PortA. Figures 2-45 to 2-47 show descriptions of these registers.

The PortA pins have alternative functions related to the optional external data SRAM. PortA can be configured to be the multiplexed low-order address/data bus during accesses to the external data memory. In this mode, PortA has internal pull-ups.

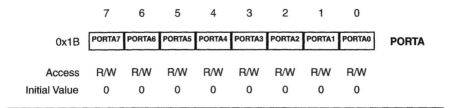

Figure 2-45
PortA Data register (PORTA).

	7	6	5	4	3	2	1	0	
0x1A	DDA7	DDA6	DDA5	DDA4	DDA3	DDA2	DDA1	DDA0	**DDRA**
Access	R/W	R/W	R/W	R/W	R/W	R/W	R/W	R/W	
Initial Value	0	0	0	0	0	0	0	0	

Figure 2-46
PortA Data Direction register (DDRA).

	7	6	5	4	3	2	1	0	
0x19	PINA7	PINA6	PINA5	PINA4	PINA3	PINA2	PINA1	PINA0	**PINA**
Access	R	R	R	R	R	R	R	R	
Initial Value	Hi-Z	Hi-Z	Hi-Z	Hi-Z	Hi-Z	Hi-Z	Hi-Z	Hi-Z	

Figure 2-47
PortA Input Pins Address (PINA).

PortA also receives the code bytes during flash programming and outputs the code bytes during program verification. External pull-ups are required during program verification.

When PortA is set to the alternative function by the External SRAM Enable bit (SRE) in the MCU control register (MCUCR), the alternative settings override the data direction register.

PortB   Three data memory address locations are allocated for PortB. Figures 2-48 to 2-50 show a description of these registers.

The PortB pins have alternative functions, which were listed in Tables 2-11 and Table 2-12 and are described in special sections.

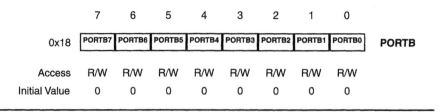

	7	6	5	4	3	2	1	0	
0x18	PORTB7	PORTB6	PORTB5	PORTB4	PORTB3	PORTB2	PORTB1	PORTB0	**PORTB**
Access	R/W	R/W	R/W	R/W	R/W	R/W	R/W	R/W	
Initial Value	0	0	0	0	0	0	0	0	

Figure 2-48
PortB Data register (PORTB).

	7	6	5	4	3	2	1	0	
0x17	DDB7	DDB6	DDB5	DDB4	DDB3	DDB2	DDB1	DDB0	DDRB
Access	R/W	R/W	R/W	R/W	R/W	R/W	R/W	R/W	
Initial Value	0	0	0	0	0	0	0	0	

Figure 2-49
PortB Data Direction register (DDRB).

	7	6	5	4	3	2	1	0	
0x16	PINB7	PINB6	PINB5	PINB4	PINB3	PINB2	PINB1	PINB0	PINB
Access	R	R	R	R	R	R	R	R	
Initial Value	Hi-Z	Hi-Z	Hi-Z	Hi-Z	Hi-Z	Hi-Z	Hi-Z	Hi-Z	

Figure 2-50
PortB Input Pins Address (PINB).

PortC (AT90S4414 and AT90S8515 Only)   Three data memory address locations are allocated for PortC. Figures 2-51 to 2-53 show descriptions of these registers.

The PortC pins have alternative functions related to the optional external data SRAM. PortC can be configured to be the high-order address byte during access to external data memory. In this mode, PortC uses internal pull-ups when emitting ones.

PortC also receives the high-order address bits and some control signals during flash programming and verification.

When PortC is set to the alternative function by the External SRAM Enable bit (SRE) in the MCU control register, the alternative settings override the data direction register.

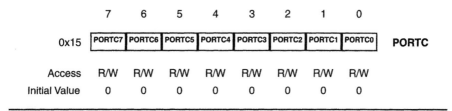

	7	6	5	4	3	2	1	0	
0x15	PORTC7	PORTC6	PORTC5	PORTC4	PORTC3	PORTC2	PORTC1	PORTC0	PORTC
Access	R/W	R/W	R/W	R/W	R/W	R/W	R/W	R/W	
Initial Value	0	0	0	0	0	0	0	0	

Figure 2-51
PortC Data register (PORTC).

7	6	5	4	3	2	1	0	
0x14	DDC7	DDC6	DDC5	DDC4	DDC3	DDC2	DDC1	DDC0	DDRC

Access	R/W	R/W	R/W	R/W	R/W	R/W	R/W	R/W
Initial Value	0	0	0	0	0	0	0	0

Figure 2-52
PortC Data Direction register (DDRC).

	7	6	5	4	3	2	1	0	
0x13	PINC7	PINC6	PINC5	PINC4	PINC3	PINC2	PINC1	PINC0	PINC

Access	R	R	R	R	R	R	R	R
Initial Value	Hi-Z	Hi-Z	Hi-Z	Hi-Z	Hi-Z	Hi-Z	Hi-Z	Hi-Z

Figure 2-53
PortC Input Pins Address (PINC).

PortD   Three data memory address locations are allocated for PortD. Figures 2-54 to 2-56 show descriptions of these registers.

The PortD pins have alternative functions. These alternative functions were listed in Tables 2-11 and Table 2-12 and described in special sections.

Access to External SRAM   The access to external SRAM is supported by the alternative functions of PortA and PortC. PortA serves as a multiplexed low-order address bus (A0 to A7) or data bus (D0 to D7), respectively. The output ALE signalizes the state of the multiplexer.

With the falling edge of ALE, the valid low-order addresses on PortA are saved in a byte-wide latch. The outputs of this latch are connected to the low-

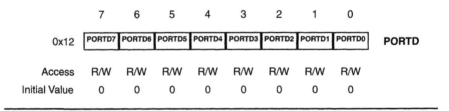

	7	6	5	4	3	2	1	0	
0x12	PORTD7	PORTD6	PORTD5	PORTD4	PORTD3	PORTD2	PORTD1	PORTD0	PORTD

Access	R/W	R/W	R/W	R/W	R/W	R/W	R/W	R/W
Initial Value	0	0	0	0	0	0	0	0

Figure 2-54
PortD Data register (PORTD).

AVR RISC Microcontroller Handbook

	7	6	5	4	3	2	1	0	
0x11	DDD7	DDD6	DDD5	DDD4	DDD3	DDD2	DDD1	DDD0	**DDRD**
Access	R/W	R/W	R/W	R/W	R/W	R/W	R/W	R/W	
Initial Value	0	0	0	0	0	0	0	0	

Figure 2-55
PortD Data Direction register (DDRD).

	7	6	5	4	3	2	1	0	
0x10	PIND7	PIND6	PIND5	PIND4	PIND3	PIND2	PIND1	PIND0	**PIND**
Access	R	R	R	R	R	R	R	R	
Initial Value	Hi-Z	Hi-Z	Hi-Z	Hi-Z	Hi-Z	Hi-Z	Hi-Z	Hi-Z	

Figure 2-56
PortD Input Pins Address (PIND).

order address pins of the external memory. The high-order addresses are sta-ble on PortC from the beginning of that memory access, before the read strobe (/RD) or write strobe (/WR) occurs, so all address lines are stable for addressing external memory.

Reading data from external memory, the rising edge of the read strobe writes the data to PortA. Figure 2-57 shows the external memory read cycle.

The rising edge of the write strobe writes the data to the external memory. Figure 2-58 shows the external memory write cycle.

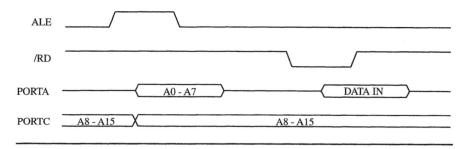

Figure 2-57
Reading external SRAM.

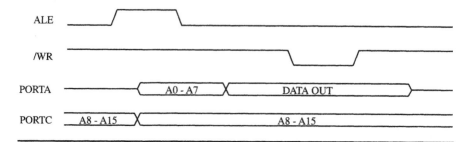

Figure 2-58
Writing external SRAM.

The required circuitry for connecting external SRAM to AVR microcontrollers is quite simple, as Figure 2-59 shows. A 74HCT573 latch saves the low-order address bits controlled by ALE. The storage capacity of the used SRAM is 32K × 8. Therefore, only the address bits A8 to A14 are connected to PortC directly. Address bit A15 is used as chip select /CS. Therefore the SRAM will be activated only for accesses to addresses between 0000_H and $7FFF_H$. The control inputs /OE and /WE of the used SRAM serve as read strobe (/OE) respectively write strobe (/WE).

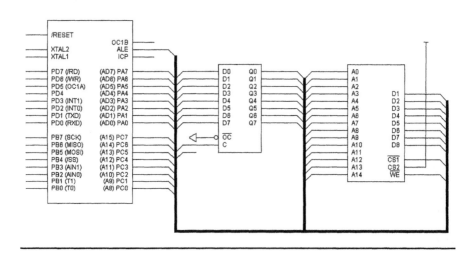

Figure 2-59
Connection of external SRAM to AT90S8515.

2.5 Reset and Interrupt System

The microcontrollers of the AVR family have one reset and can handle as many as 12 different interrupt sources.

Each interrupt can be enabled by an individual enable bit. To enable an interrupt, the necessary enable bit and the Global Interrupt Enable bit in the status register must be set together.

Table 2-14 shows the available resources for the related AVR microcontrollers.

2.5.1 Interrupt Vector Table

These interrupts and the separate reset vector each have a separate program vector in the program memory space. The lowest addresses in the program memory space are automatically defined as the reset and interrupt vectors.

The order of the interrupt vectors determines the priority levels of the different interrupts. The lower the address, the higher the priority level. RESET has the highest priority; next is INT0, the External Interrupt Request 0; and so on.

Table 2-14

Resets and interrupts on AVR microcontrollers.

Reset/Interrupt Source	Reset / Interrupt Definition	S1200	S2313	S4414	S8515
RESET	Hardware Pin and Watchdog Reset	☒	☒	☒	☒
INT0	External Interrupt Request 0	☒	☒	☒	☒
INT1	External Interrupt Request 1		☒	☒	☒
TIMER1 CAPT	Timer/Counter1 Capture Event		☒	☒	☒
TIMER1 COMPA	Timer/Counter1 Compare Match A		☒	☒	☒
TIMER1 COMPB	Timer/Counter1 Compare Match B			☒	☒
TIMER1 OVF	Timer/Counter1 Overflow		☒	☒	☒
TIMER0 OVF	Timer/Counter0 Overflow	☒	☒	☒	☒
SPI STC	Serial Transfer Complete			☒	☒
UART RX	UART Rx Complete		☒	☒	☒
UART UDRE	UART Data Register Empty		☒	☒	☒
UART TX	UART Tx Complete		☒	☒	☒
ANA_COMP	Analog Comparator	☒	☒	☒	☒

The complete interrupt vector tables for all microcontrollers of the AVR family are shown in Tables 2-15, 2-16, and 2-17.

Note that the order of interrupts is not the same for the different models of AVR microcontrollers. When software written for one AVR microcontroller is ported to another member of the AVR family, it is essential to consider these possibly changed locations.

2.5.2 Reset Sources

Three different sources can reset the AVR microcontrollers. Table 2-18 lists these reset sources.

Table 2-15
AT90S1200 interrupt vector table.

Program Address	Source	Interrupt Definition
0x00	RESET	/RESET and Watchdog Reset
0x01	INT0	External Interrupt Request 0
0x02	TIMER0 OVF	Timer/Counter0 Overflow
0x03	ANA_COMP	Analog Comparator

Table 2-16
AT90S2312 interrupt vector table.

Program Address	Source	Interrupt Definition
0x00	RESET	/RESET and Watchdog Reset
0x01	INT0	External Interrupt Request 0
0x02	INT1	External Interrupt Request 1
0x03	TIMER1 CAPT	Timer/Counter1 Capture Event
0x04	TIMER1 COMP	Timer/Counter1 Compare Match A
0x05	TIMER1 OVF	Timer/Counter1 Overflow
0x06	TIMER0 OVF	Timer/Counter0 Overflow
0x07	UART, RX	UART Receive Complete
0x08	UART, UDRE	UART Data Register Empty
0x09	UART, TX	UART Transmit Complete
0x0A	ANA_COMP	Analog Comparator

Table 2-17
Interrupt vector table for AT90S4414 and AT90S8515.

Program Address	Source	Interrupt Definition
0x00	RESET	/RESET and Watchdog Reset
0x01	INT0	External Interrupt Request 0
0x02	INT1	External Interrupt Request 1
0x03	TIMER1 CAPT	Timer/Counter1 Capture Event
0x04	TIMER1 COMPA	Timer/Counter1 Compare Match A
0x05	TIMER1 COMPB	Timer/Counter1 Compare Match B
0x06	TIMER1 OVF	Timer/Counter1 Overflow
0x07	TIMER0 OVF	Timer/Counter0 Overflow
0x09	SPI, STC	SPI Serial Transfer Complete
0x0A	UART, RX	UART Receive Complete
0x0B	UART, UDRE	UART Data Register Empty
0x0C	UART, TX	UART Transmit Complete
0x0D	ANA_COMP	Analog Comparator

Power-On Reset The power-on reset avoids the start of the operation of the microcontroller before the supply voltage has reached a safe level. After supply voltage has reached the power-on reset threshold, typically 2 V, the internal power-on reset circuitry resets for typically 3 ms, followed by a reset delay time-out period of typically 16 ms before the microcontroller starts program execution.

The /RESET pin is pulled up internally by a resistor between 10 ohm and 50 ohm. If no external reset is required, this pin can be left unconnected. Connecting the /RESET pin to V_{CC} directly has the same effect. Holding the /RESET-pin low for a period after supply voltage has been applied extends the power-on reset period.

Table 2-18
Reset sources for AVR microcontrollers.

Reset Source	The Microcontroller Is Reset:
Power-on reset	After applying a supply voltage to V_{CC} and GND pin.
External reset	When a low level is present on the /RESET pin for more than two XTAL cycles.
Watchdog reset	When the watchdog timer period expires and the watchdog was enabled before.

In cases where power supply decreases below the power-on reset threshold during operation of the AT90S1200 and AT90S8515, power-monitoring devices must be added. Later AVR microcontrollers will have a brown-out detector that resets the device on decreasing voltage as well.

External Reset An external reset is generated by a low level on the /RESET pin. This pin must be held low for at least two crystal clock cycles. When the voltage on the /RESET pin reaches the reset threshold voltage of about $V_{CC}/2$ on its rising edge, the delay timer starts the microcontroller after the reset delay time-out period of about 16 ms has expired.

To overcome the drawback of missing brown-out circuitry, it is possible to connect a power-monitoring device controlling the external reset pin. As an example of such a device, the operation of the DS1233 EconoReset monitor from Dallas is explained here. This device monitors two vital conditions for a microcontroller: power supply and external override.

Precision temperature-compensated reference and comparator circuits are used to monitor the status of the power supply. When an out-of-tolerance condition is detected, the reset output will be switched to Lo. When the supply voltage returns to an in-tolerance condition, the reset signal is kept in the Lo state for approximately 350 ms to allow the power supply and microcontroller to stabilize.

The second function of the DS1233 is push-button reset control. The device debounces a push-button closure and will generate a 350-ms reset pulse upon release. For proper operation with an external push-button, a capacitor between 100 pF and 0.01 µF must be connected between /RST and GND. Figure 2-60 shows the connection of the DS1233 to an AT90S1200.

Watchdog Reset When the watchdog times out, it will generate a short reset pulse of 1 XTAL cycle duration. On the falling edge of this pulse, the delay timer starts counting the reset delay time-out period, typically 16 ms.

For all other watchdog-related parameters refer to section 2.4.2.

2.6 Clock

For independent clock generation, either a quartz crystal or a ceramic resonator may be used. The pins XTAL1 and XTAL2 are input and output, respectively, of an inverting amplifier that can be configured for use as an on-chip oscillator. Figure 2-61 shows all required parts of the oscillator circuitry and their connections. The oscillator connections are the same for all AVR microcontrollers.

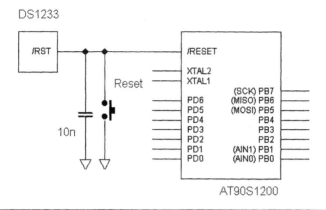

Figure 2-60
Power monitoring and push-button reset with DS1233.

In some cases a microcontroller environment has external clock sources already. It is possible to drive the microcontrollers of the AVR family by an external clock as well.

To drive the device from an external clock source, XTAL2 should be left unconnected while XTAL1 is connected to the external clock source. Figure 2-62 shows the needed connections.

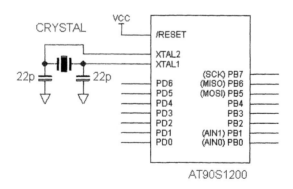

Figure 2-61
Oscillator connections.

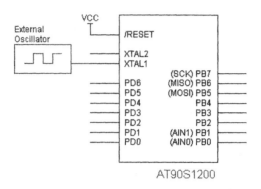

Figure 2-62
External clock source.

The AT90S1200 contains an on-chip RC oscillator in addition. This on-chip RC oscillator, running at a fixed frequency of 1 MHz, can be selected as the clock source for this microcontroller. If enabled, the AT90S1200 can operate with no external components.

The on-chip RC oscillator is selected as the clock source by the control bit RCEN in the flash memory. The RCEN bit can be changed only by parallel programming. When the on-chip RC oscillator is used for serial program downloading, the RCEN bit must first be set by parallel programming.

AVR RISC Microcontroller Handbook

Handling the Hardware Resources   3

3.1 Memory Addressing Modes

The microcontrollers of the AVR family support powerful and efficient addressing modes for access to the program memory (flash) and data memory (SRAM). This chapter describes the different addressing modes supported by the AVR architecture.

3.1.1 Register Direct Addressing

The immediate instructions include their operands in the instruction word and access directly to the destination register, as Figure 3-1 shows.

An example of direct single register addressing is the instruction ldi r0,$FF. The immediate value $FF will be stored in register r0.

Many instructions contain the addresses of both operands in their instruction word. We speak about direct addressing of two registers. Figure 3-2 shows this addressing mode.

An example of direct register addressing of two registers is the instruction add r0,r1. The contents of the registers r0 and r1 will be added and the result will be stored in register r0.

3.1.2 I/O Direct Addressing

Instruction that handle the I/O memory contain both the port address (6 bit) and either the source or destination register (5 bit) of that operation, as shown in Figure 3-3.

An example of I/O direct register addressing is the instruction out PORTD,r0. The contents of the register r0 will be stored in the I/O port PORTD (= $12).

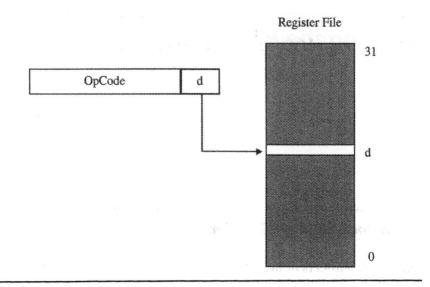

Figure 3-1
Direct single-register addressing.

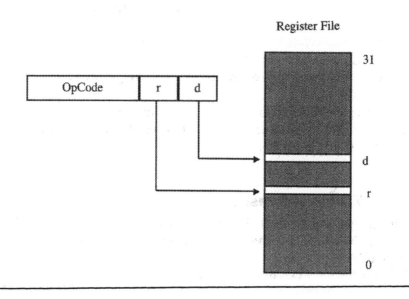

Figure 3-2
Direct register addressing, two registers.

AVR RISC Microcontroller Handbook

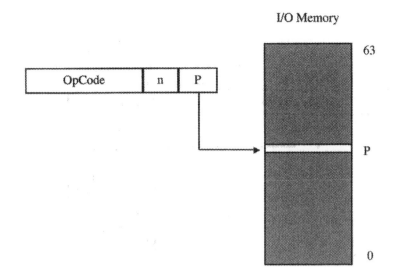

Figure 3-3
I/O Direct addressing.

3.1.3 SRAM Direct Addressing

In direct addressing of SRAM, the SRAM address is contained in the lower word of this two-word instruction. The source or destination register (5 bit) of that operation is contained in the higher word. Figure 3-4 shows this addressing mode.

An example of SRAM direct register addressing is the instruction lds r0,$1000. Register r0 will be loaded with the contents of SRAM location $1000.

3.1.4 SRAM Indirect Addressing

The indirect data addressing modes are very powerful features of the AVR architecture. The overhead to access large data arrays stored in SRAM can be minimized with these modes.

As Figure 3-5 shows, for indirect data addressing the operand address is the contents of the X-, Y-, or Z-register. The source or destination register (5 bit) of that operation is contained in the instruction word.

An example of indirect SRAM addressing is the instruction ld r4,z. Register r4 will be loaded with the contents of the SRAM location defined by the contents of the Z-register.

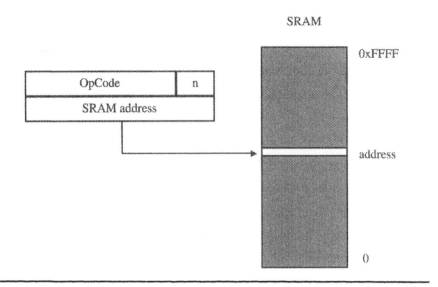

Figure 3-4
Direct SRAM addressing.

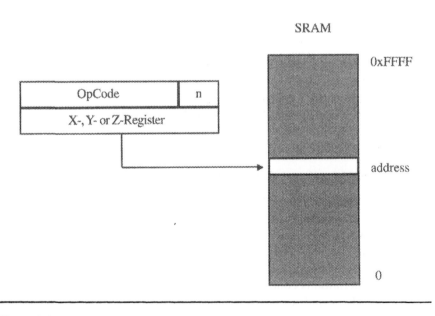

Figure 3-5
SRAM indirect addressing.

For indirect data addressing with displacement the operand address results from the contents of the Y- or Z-register added to a displacement value (6 bits) contained in the instruction word. The source or destination register (5 bit) of that operation is also contained in the instruction word. Figure 3-6 shows the required calculation before addressing SRAM.

An example of indirect SRAM addressing is the instruction ldd r4,Y+2. Register r4 will be loaded with the contents of the SRAM location defined by the contents of the Y-register incremented by 2.

For indirect data addressing with pre-decrement, the operand address results from the contents of the X-, Y-, or Z-register decremented before addressing SRAM. The original contents of the X-, Y-, or Z-register is replaced by the decremented value after that operation. The source or destination register (5 bit) of that operation is contained in the instruction word. Figure 3-7 shows the required calculation before addressing SRAM.

An example of indirect SRAM addressing with pre-decrement is the instruction ldd r4,Y-. Register r4 will be loaded with the contents of the SRAM location defined by the contents of the Y-register decremented by 1. After that operation, the Y-register contains the decremented value.

For indirect data addressing with post-increment, the operand address is the contents of the X-, Y-, or Z-register. After that operation the contents of the X-, Y-, or Z-register are incremented by 1. The source or destination register (5 bit) of that operation is contained in the instruction word. Figure 3-8 shows this addressing mode.

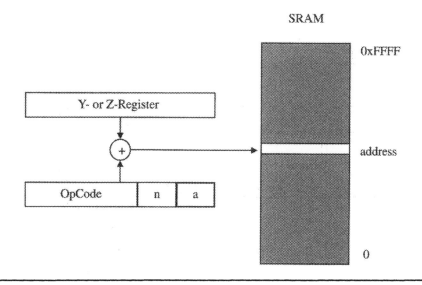

Figure 3-6
SRAM indirect addressing with displacement.

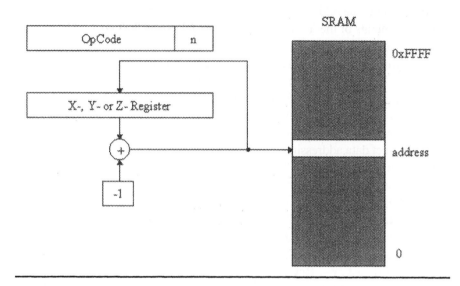

Figure 3-7
SRAM indirect addressing with pre-decrement.

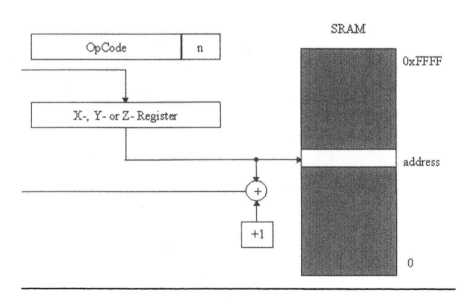

Figure 3-8
SRAM indirect addressing with post-increment.

AVR RISC Microcontroller Handbook

An example of indirect SRAM addressing with post-increment is the instruction ldd r4,X+. Register r4 will be loaded with the contents of the SRAM location addressed by the contents of the X-register. After that operation, the X-register contains the incremented address value for the next operation.

3.1.5 Constant Addressing Using the LPM Instruction

Access to tables stored in program memory can be handled very easy by the LPM instruction. The LPM instruction loads 1 byte indirectly from program memory to a register.

The program memory location is pointed to by the Z (16 bit) pointer register in the register file. Memory access is limited to 4K words. The LSB of the Z register distinguishes between Hi (LSB = 1) and Lo byte (LSB = 0). Figure 3-9 shows addressing program memory with instruction lpm.

3.1.6 Jumps and Calls

For changing the order of instruction execution—this means branching or subroutine calls—different jump and call instructions are available. We have to distinguish among direct, indirect, and relative jumps and calls.

Figure 3-10 shows direct memory addressing in jmp and call instructions, named long jump or long call, respectively.

For the indirect addressing used in the ijmp and icall instuctions, the Z-register contains the memory location. Figure 3-11 shows this memory access.

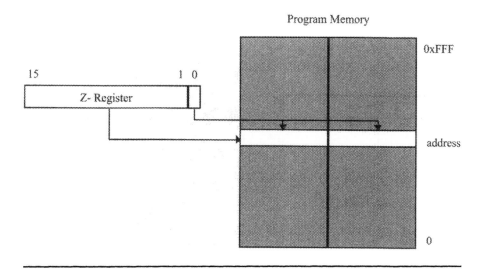

Figure 3-9
Program memory constant addressing.

Program Memory

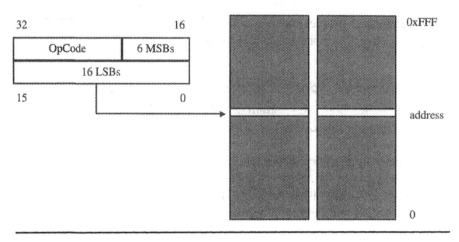

Figure 3-10
Direct program memory addressing.

The relative addressing used in `rjmp` and `rcall` instructions is effective for jumps or calls in a reduced memory area. The jumps or calls work relative to the program counter (PC). Figure 3-12 shows the memory addressing for the `rjmp` and `rcall` instructions.

For AVR microcontrollers with program memory not exceeding 4K words (8K bytes), this instruction can address the entire memory from every address location.

Program Memory

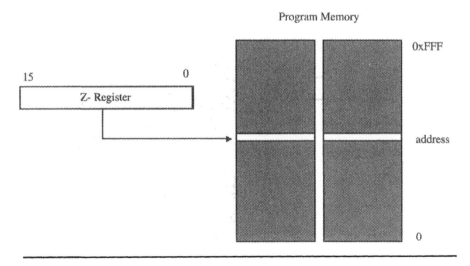

Figure 3-11
Indirect program memory addressing.

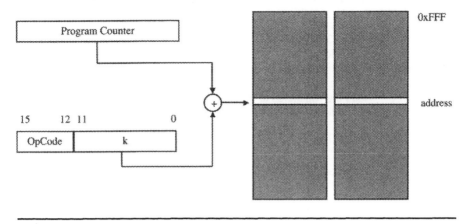

Figure 3-12
Relative program memory addressing.

3.2 Instruction Set

In this section the whole instruction set for the first four microcontrollers of the AVR family—AT90S1200, AT90S2313, AT90S4414, and AT90S8515—is explained in detail. Not all instructions are implemented in each microcontroller of the AVR family. Therefore, the head of each instruction description shows the devices concerned.

ADC	*Add with Carry*	*AT90S1200/2313/4414/8515*

Syntax	`ADC Rd,Rr`
Operands	$0 \leq d \leq 31, 0 \leq r \leq 31$
Operation	Rd ← Rd + Rr + C
Flags Affected	H, S, V, N, Z, C
Encoding	`0001 11rd dddd rrrr`
Description	Adds the registers Rd and Rr and the contents of the C flag and places the result in the destination register Rd.
Words	1
Cycles	1

Example	`            ; Add R1:R0 to R3:R2`
	`add r2,r0 ; Add low bytes`
	`add r3,r1 ; Add high bytes with carry`

ADD	Add without Carry	AT90S1200/2313/4414/8515

Syntax	ADD Rd,Rr
Operands	$0 \le d \le 31, 0 \le r \le 31$
Operation	Rd ← Rd + Rr
Flags Affected	H, S, V, N, Z, C
Encoding	0000 11rd dddd rrrr
Description	Adds the registers Rd and Rr and places the result in the destination register Rd.
Words	1
Cycles	1
Example	add r1,r2 ; Add r2 to r1
	add r28,r28 ; Add r28 to itself

ADIW	Add Immediate to Word	AT90S2313/4414/8515

Syntax	ADIW Rdl,K
Operands	dl ∈ {24,26,28,30}, $0 \le K \le 63$
Operation	Rdh:Rdl ← Rdh:Rdl + K
Flags Affected	S, V, N, Z, C
Encoding	1001 0110 KKdd KKKK
Description	Adds an immediate value (0–63) to a register pair and places the result in the register pair. This instruction operates on the upper four register pairs and is well suited for operations on the pointer registers.
Words	1
Cycles	2
Example	adiw r24,1 ; Add 1 to r25:r24
	adiw r30,63 ; Add 63 to the Z pointer(r31:r30)

AND	Logical AND	AT90S1200/2313/4414/8515

Syntax	AND Rd,Rr
Operands	$0 \le d \le 31, 0 \le r \le 31$
Operation	Rd ← Rd AND Rr
Flags Affected	S, V, N, Z
Encoding	0010 00rd dddd rrrr
Description	Performs a logical AND between the contents of registers Rd and Rr and places the result in the destination register Rd.
Words	1
Cycles	1
Example	and r2,r3 ; Bitwise AND of r2 and r3, result in r2
	ldi r16,1 ; Load 0000 0001 to r16 immediately
	and r2,r16 ; Isolate bit0 in r2

ANDI	*Logical AND with Immediate*	*AT90S1200/2313/4414/8515*

Syntax	`ANDI Rd,K`
Operands	$16 \leq d \leq 31$, $0 \leq K \leq 255$
Operation	Rd ← Rd AND K
Flags Affected	S, V, N, Z
Encoding	`0111 KKKK dddd KKKK`
Description	Performs a logical AND between the contents of registers Rd and a byte constant and places the result in the destination register Rd.
Words	1
Cycles	1
Example	`andi r17,$0F ; Clear upper nibble of r17`
	`andi r18,$10 ; Isolate bit4 in r18`
	`andi r19,$AA ; Clear odd bits of r19`

ASR	*Arithmetic Shlft Right*	*AT90S1200/2313/4414/8515*

Syntax	`ASR Rd,Rr`
Operands	$0 \leq d \leq 31$
Operation	

B7 B6 B5 B4 B3 B2 B1 B0

Flags Affected	S, V, N, Z, C
Encoding	`1001 010d dddd 0101`
Description	Shifts all bits in register Rd one place to the right. Bit7 is held constant and bit0 is loaded into the C flag.
Words	1
Cycles	1
Example	`ldi r16,$10 ; Load 16`$_D$` into r16`
	`asr r16    ; r16 = r16/2`
	`ldi r17,$FC ; Load -4`$_D$` into r17`
	`asr r17    ; r17 = r17/2`

BCLR	*Bit Clear in SREG*	*AT90S1200/2313/4414/8515*

Syntax	`BCLR s`
Operands	$0 \leq s \leq 7$
Operation	SREG(s) ← 0
Flags Affected	I, T, H, S, V, N, Z, C
Encoding	`1001 0100 1sss 1000`
Description	Clears a single flag in SREG.
Words	1
Cycles	1
Example	`bclr 0 ; Clears bit 0 in SREG`
	`bclr 7 ; Disables interrupts`

BLD	Bit Load from T Flag in SREG to a Bit in Register	AT90S1200/2313/4414/8515

Syntax	BLD Rd,b
Operands	$0 \leq d \leq 31, 0 \leq b \leq 7$
Operation	Rd(b) ← T
Flags Affected	none
Encoding	1111 100d dddd Xbbb
Description	Copies the T flag in SREG to bit b in register Rd.
Words	1
Cycles	1
Example	bst r1,2 ; Store bit2 if r1 in T flag
	bld r0,4 ; Load T flag in bit4 of r0

BRBC	Branch If Bit in SREG Is Cleared	AT90S1200/2313/4414/8515

Syntax	BRBC s,k
Operands	$0 \leq s \leq 7, -64 \leq k \leq 63$
Operation	if SREG(s) = 0 then branch k steps
Flags Affected	none
Encoding	1111 01kk kkkk ksss
Description	Tests a single bit in SREG and branches relatively to PC if the bit is cleared. The parameter k is the offset from PC and is represented in two's-complement form.
Words	1
Cycles	1 (false), 2 (true)
Example	cpi r20,5 ; Compare r20 to the value 5
	brbc 1,noteq ; Branch to label noteq,
	; if zero flag is cleared

BRBS	Branch If Bit in SREG Is Set	AT90S1200/2313/4414/8515

Syntax	BRBS s,k
Operands	$0 \leq s \leq 7, -64 \leq k \leq 63$
Operation	if SREG(s) = 1 then branch k steps
Flags Affected	none
Encoding	1111 00kk kkkk ksss
Description	Tests a single bit in SREG and branches relatively to PC if the bit is set. The parameter k is the offset from PC and is represented in two's-complement form.
Words	1
Cycles	1 (true), 2 (false)
Example	bst r0,3 ; Load T flag with bit3 of r0
	brbs 6,bitset ; Branch to label bitset,
	; if T flag was set

BRCC	*Branch If Carry Is Cleared*	*AT90S1200/2313/4414/8515*

Syntax	`BRCC k`
Operands	-64 ≤ k ≤ 63
Operation	if C = 0 then branch k steps
Flags Affected	none
Encoding	`1111 01kk kkkk k000`
Description	Tests the Carry flag (C) and branches relatively to PC if C is cleared. The parameter k is the offset from PC and is represented in two's-complement form (equivalent to BRBC 0,k).
Words	1
Cycles	1 (false), 2 (true)
Example	`add r22,r23  ; Add r23 to r22`
	`brcc nocarry ; Branch to label nocarry,`
	`             ; if carry flag is cleared`

BRCS	*Branch If Carry Is Set*	*AT90S1200/2313/4414/8515*

Syntax	`BRCS k`
Operands	-64 ≤ k ≤ 63
Operation	if C = 1 then branch k steps
Flags Affected	none
Encoding	`1111 00kk kkkk k000`
Description	Tests the Carry flag (C) and branches relatively to PC if C is set. The parameter k is the offset from PC and is represented in two's-complement form (equivalent to BRBS 0,k).
Words	1
Cycles	1 (false), 2 (true)
Example	`cpi r26,$56 ; Compare r26 with value $56`
	`brcs carry  ; Branch to label carry,`
	`            ; if carry flag is set`

BREQ	*Branch If Equal*	*AT90S1200/2313/4414/8515*

Syntax	`BREQ k`
Operands	-64 ≤ k ≤ 63
Operation	if Z = 1 then branch k steps
Flags Affected	none
Encoding	`1111 00kk kkkk k001`
Description	Tests the Zero flag (Z) and branches relatively to PC if Z is set.
	If the instruction is executed immediately after any of the instructions CP, CPI, SUB, or SUBI, the branch will occur if, and only if, the unsigned or signed binary number represented in Rd was equal to the the unsigned or signed binary number represented in Rr.
	The parameter k is the offset from PC and is represented in two's-complement form (equivalent to BRBS 1,k).

BREQ	Branch If Equal	AT90S1200/2313/4414/8515

Words   1
Cycles   1 (false), 2 (true)
Example
```
cp r1,r0   ; Compare r1 and r0
breq equal ; Branch to label equal,
           ; if registers are equal
           ; (e.g., zero flag is set)
```

	Branch If Greater or	
BRGE	Equal (Signed)	AT90S1200/2313/4414/8515

Syntax   BRGE k
Operands $-64 \leq k \leq 63$
Operation   if S = 0 then branch k steps
Flags Affected none
Encoding   1111 01kk kkkk k100
Description   Tests the Signed flag (S) and branches relatively to PC if S is cleared.

If the instruction is executed immediately after any of the instructions CP, CPI, SUB, or SUBI, the branch will occur if, and only if, the signed binary number represented in Rd was greater than, or equal to, the signed binary number represented in Rr.

The parameter k is the offset from PC and is represented in two's-complement form (equivalent to BRBS 1,k).

Words   1
Cycles   1 (false), 2 (true)
Example
```
cp r11,r12   ; Compare r11 and r12
brge greateq ; Branch to label greateq,
             ; if r11 ≥ r12
             ; (e.g., signed flag is cleared)
```

	Branch If Half Carry	
BRHC	Flag Is Cleared	AT90S1200/2313/4414/8515

Syntax   BRHC k
Operands $-64 \leq k \leq 63$
Operation   if H = 0 then branch k steps
Flags Affected none
Encoding   1111 01kk kkkk k101
Description   Tests the Half Carry flag (H) and branches relatively to PC of H is cleared. The parameter k is the offset from PC and is represented in two's-complement form (equivalent to instruction BRBC 5,k).

Words   1
Cycles   1 (false), 2 (true)

Example	`brhc hclear ; Branch if half carry flag is`
	`                     ; cleared`
	`       . . .`
	`hclear: nop         ; Branch destination (do nothing)`

BRHS	Branch If Half Carry Flag Is Set	AT90S1200/2313/4414/8515

Syntax	`BRHS k`
Operands	$-64 \le k \le 63$
Operation	if H = 1 then branch k steps
Flags Affected	none
Encoding	`1111 00kk kkkk k101`
Description	Tests the Half Carry flag (H) and branches relatively to PC if H is set. The parameter k is the offset from PC and is represented in two's-complement form (equivalent to instruction BRBS 5,k).
Words	1
Cycles	1 (false), 2 (true)
Example	`brhs hset ; Branch if half carry flag is set`
	`      . . .`
	`hset: nop       ; Branch destination (do nothing)`

BRID	Branch If Global Interrupt Is Disabled	AT90S1200/2313/4414/8515

Syntax	`BRID k`
Operands	$-64 \le k \le 63$
Operation	if I = 0 then branch k steps
Flags Affected	none
Encoding	`1111 01kk kkkk k111`
Description	Tests the Global Interrupt flag (I) and branches relatively to PC if I is cleared. The parameter k is the offset from PC and is represented in two's-complement form (equivalent to instruction BRBC 7,k).
Words	1
Cycles	1 (false), 2 (true)
Example	`brid intdis ; Branch if interrupt disabled`
	`        . . .`
	`intdis: nop          ; Branch destination`

BRIE	Branch If Global Interrupt Is Enabled	AT90S1200/2313/4414/8515

Syntax	`BRIE k`
Operands	$-64 \le k \le 63$
Operation	if I = 1 then branch k steps

BRIE	Branch If Global Interrupt Is Enabled	AT90S1200/2313/4414/8515

Flags Affected	none
Encoding	`1111 00kk kkkk k111`
Description	Tests the Global Interrupt flag (I) and branches relatively to PC if I is set. The parameter k is the offset from PC and is represented in two's-complement form (equivalent to instruction BRBS 7,k).
Words	1
Cycles	1 (false), 2 (true)
Example	

```
        brid inten ; Branch if interrupt enabled
        . . .
inten: nop        ; Branch destination
```

BRLO	Branch If Lower	AT90S1200/2313/4414/8515

Syntax	`BRLO k`
Operands	$-64 \leq k \leq 63$
Operation	if C = 1 then branch k steps
Flags Affected	none
Encoding	`1111 00kk kkkk k000`
Description	Tests the Carry flag (C) and branches relatively to PC if C is set.
	If the instruction is executed immediately after any of the instructions CP, CPI, SUB, or SUBI, the branch will occur if, and only if, the unsigned binary number represented in Rd was smaller than the unsigned binary number represented in Rr.
	The parameter k is the offset from PC and is represented in two's-complement form (equivalent to instruction BRBS 0,k).
Words	1
Cycles	1 (false), 2 (true)
Example	

```
        eor r19,r19 ; Clear r19
loop: inc r19       ; Increase r19
        . . .
        cpi r19,$10 ; Compare r19 with $10
        brlo loop   ; Branch if r19 < $10 (unsigned)
        nop         ; Exit from loop (do nothing)
```

BRLT	Branch If Less Than (Signed)	AT90S1200/2313/4414/8515

Syntax	`BRLT k`
Operands	$-64 \leq k \leq 63$
Operation	if S = 1 then branch k steps
Flags Affected	none

Encoding	`1111 00kk kkkk k100`
Description	Tests the Sign flag (S) and branches relatively to PC if S is set. The parameter k is the offset from PC and is represented in two's-complement form (equivalent to instruction BRBS 2,k).
Words	1
Cycles	1 (false), 2 (true)
Example	

```
        cp r16,r1 ; Compare r16 to r1
        brlt less ; Branch if r16 < r1 (signed)
        ...
less: nop ; Branch destination (do nothing)
```

BRMI	*Branch If Minus*	*AT90S1200/2313/4414/8515*

Syntax	`BRMI k`
Operands	$-64 \leq k \leq 63$
Operation	if N = 1 then branch k steps
Flags Affected	none
Encoding	`1111 00kk kkkk k010`
Description	Tests the Negative flag (N) and branches relatively to PC if N is set. This instruction branches relatively to PC in either. The parameter k is the offset from PC and is represented in two's-complement form (equivalent to instruction BRBS 2,k).
Words	1
Cycles	1 (false), 2 (true)
Example	

```
        subi r18,4 ; Subtract 4 from r18
        brmi neg   ; Branch if result negative
        ...
neg: nop           ; Branch destination (do nothing)
```

BRNE	*Branch If Not Equal*	*AT90S1200/2313/4414/8515*

Syntax	`BRNE k`
Operands	$-64 \leq k \leq 63$
Operation	if Z = 0 then branch k steps
Flags Affected	none
Encoding	`1111 01kk kkkk k001`
Description	Tests the Zero flag (Z) and branches relatively to PC if Z is cleared.

If the instruction is executed immediately after any of the instructions CP, CPI, SUB, or SUBI, the branch will occur if, and only if, the unsigned or signed binary number represented in Rd was not equal to the unsigned or signed binary number represented in Rr.

The parameter k is the offset from PC and is represented in two's-complement form (equivalent to instruction BRBC 1,k).

BRNE	Branch If Not Equal	AT90S1200/2313/4414/8515

Words   1
Cycles   1 (false), 2 (true)
Example

```
            eor r27,r27 ; Clear r27
    loop: inc r27      ; Increase r27
            . . .
            cpi r27,5   ; Compare r27 to 5
            brne loop   ; Branch if r27<>5
            nop         ; Loop exit (do nothing)
```

BRPL	Branch If Plus	AT90S1200/2313/4414/8515

Syntax   BRPL k
Operands $-64 \leq k \leq 63$
Operation   if N = 0 then branch k steps
Flags Affected none
Encoding   1111 01kk kkkk k010
Description Tests the Negative flag (N) and branches relatively to PC if N is cleared. The parameter k is the offset from PC and is represented in two's-complement form (equivalent to instruction BRBC 2,k).
Words   1
Cycles   1 (false), 2 (true)
Example

```
            subi r26,$50  ; Subtract $50 from r26
            brpl positive ; Branch if r26 positive
            . . .
    positive: nop         ; Branch destination
```

	Branch If Same or	
BRSH	Higher (Unsigned)	AT90S1200/2313/4414/8515

Syntax   BRSH k
Operands $-64 \leq k \leq 63$
Operation   if C = 0 then branch k steps
Flags Affected none
Encoding   1111 01kk kkkk k000
Description Tests the Carry flag (C) and branches relatively to PC if C is cleared.

If the instruction is executed immediately after any if the instructions CP, CPI, SUB, or SUBI, the branch will occur if, and only if, the unsigned binary number represented in Rd was greater than or equal to, the unsigned binary number represented in Rr.

The parameter k is the offset from PC and is represented in two's-complement form (equivalent to instruction BRBC 0,k).

Words	1
Cycles	1 (false), 2 (true)
Example	subi r19,4 ; Subtract 4 from r19
	brsh highsm ; Branch to label highsm,
	; if r19 ≥ 4
	; (e.g., carry flag is cleared)

BRTC	*Branch If T Flag Is Cleared*	*AT90S1200/2313/4414/8515*

Syntax	BRTC k
Operands	-64 ≤ k ≤ 63
Operation	if T = 0 then branch k steps
Flags Affected	none
Encoding	1111 01kk kkkk k100
Description	Tests the T flag and branches relatively to PC if T is cleared. The parameter k is the offset from PC and is represented in two's-complement form (equivalent to instruction BRBC 6,k).
Words	1
Cycles	1 (false), 2 (true)
Example	bst r3,5 ; Store bit 5 of r3 in T flag
	brtc tclear ; Branch if T bit was cleared
	...
	tclear: nop ; Branch destination

BRTS	*Branch If T Flag Is Set*	*AT90S1200/2313/4414/8515*

Syntax	BRTS k
Operands	-64 ≤ k ≤ 63
Operation	if T = 1 then branch k steps
Flags Affected	none
Encoding	1111 00kk kkkk k110
Description	Tests the T flag and branches relatively to PC if T is set. The parameter k is the offset from PC and is represented in two's-complement form (equivalent to instruction BRBS 6,k).
Words	1
Cycles	1 (false), 2 (true)
Example	bst r3,5 ; Store bit 5 of r3 in T flag
	brts tset ; Branch if T bit was set
	...
	tset: nop ; Branch destination (do nothing)

Handling the Hardware Resources   81

BRVC	*Branch If Overflow* *Flag Is Cleared*	*AT90S1200/2313/4414/8515*

Syntax	BRVC k
Operands	-64 ≤ k ≤ 63
Operation	if V = 0 then branch k steps
Flags Affected	none
Encoding	1111 01kk kkkk k011
Description	Tests the Overflow flag (V) and branches relatively to PC if V is cleared. The parameter k is the offset from PC and is represented in two's-complement form (equivalent to instruction BRBC 3,k).
Words	1
Cycles	1 (false), 2 (true)
Example	

```
        add r3,r4   ; Add r4 to r3
        brvc noover ; Branch if no overflow
        . . .
noover: nop         ; Branch destination
```

BRVS	*Branch If Overflow Flag Is Set*	*AT90S1200/2313/4414/8515*

Syntax	BRVS k
Operands	-64 ≤ k ≤ 63
Operation	if V = 1 then branch k steps
Flags Affected	none
Encoding	1111 00kk kkkk k011
Description	Tests the Overflow flag (V) and branches relatively to PC if V is set. The parameter k is the offset from PC and is represented in two's-complement form (equivalent to instruction BRBS 3,k).
Words	1
Cycles	1 (false), 2 (true)
Example	

```
        add r3,r4   ; Add r4 to r3
        brvs overfl ; Branch if overflow
        . . .
overfl: nop         ; Branch destination
```

BSET	*Bit Set in SREG*	*AT90S1200/2313/4414/8515*

Syntax	BSET s
Operands	0 ≤ s ≤ 7
Operation	SREG(s) ← 1
Flags Affected	Corresponding flag is set.
Encoding	1001 0100 0sss 1000
Description	Sets a single flag or bit in SREG.
Words	1
Cycles	1
Example	

```
bset 6 ; Set T flag
bset 7 ; Enable interrupt
```

BST	Bit Store from Bit in Register to T Flag in SREG	AT90S1200/2313/4414/8515

Syntax	`BST Rd,b`
Operands	$0 \leq d \leq 31, 0 \leq b \leq 7$
Operation	T ← 0 if bit b in register Rd is cleared, T ← 1 otherwise
Flags Affected	T
Encoding	`1111 101d dddd Xbbb`
Description	Stores bit b from Rd to the T flag in SREG (status register).
Words	1
Cycles	1

Example	`        ; Copy bit`
	`bst r1,2 ; Store bit 2 of r1 in T flag`
	`bld r0,4 ; Load T into bit 4 of r0`

CALL	Long Call to a Subroutine	AT90S1200/2313/4414/8515

Syntax	`CALL k`
Operands	$0 \leq k \leq 64K$
Operation	PC ← k
Flags Affected	none
Encoding	`1001 010k kkkk 111k kkkk kkkk kkkk kkkk`
Description	Calls to a subroutine within the entire program memory. The return address (to the instruction after the CALL) will be stored onto the stack. (See also RCALL.)
Words	2
Cycles	4

Example	`       mov r16,r0  ; Copy r0 to r16`
	`       call check  ; Call subroutine`
	`       nop         ; Continue (do nothing)`
	`       ...`
	`check: cpi r16,$42 ; Check if r16 has a special value`
	`       breq error  ; Branch if equal`
	`       ret         ; Return from subroutine`
	`       ...`
	`error: rjmp error  ; Infinite loop`

CBI	Clear Bit in I/O Register	AT90S1200/2313/4414/8515

Syntax	`CPI P, b`
Operands	$0 \leq P \leq 31, 0 \leq b \leq 7$
Operation	I/O(P,b) ← 0
Flags Affected	none
Encoding	`1001 1000 PPPP Pbbb`

CBI	Clear Bit in I/O Register	AT90S1200/2313/4414/8515

Description	Clears a specified bit in an I/O register. This instruction operates on the lower 32 I/O registers—addresses 0–31.
Words	1
Cycles	2
Example	`cbi $12,7 ; Clear bit 7 in Port D`

CBR	Clear Bits in Register	AT90S1200/2313/4414/8515

Syntax	CBR Rd, K
Operands	$0 \leq d \leq 31$, $0 \leq K \leq 255$
Operation	Rd ← Rd AND ($FF-K)
Flags Affected	S, V, N, Z
Encoding	`0111 KKKK dddd KKKK` K = ($FF−K)
Description	Clears the specified bits in register Rd. Performs the logical AND between the contents of register Rd and the complement of the constant mask K. The result will be placed in register Rd.
Words	1
Cycles	1
Example	`cbr r16,$F0 ; Clear upper nibble of r16` `cbr r18,1  ; Clear bit 0 in r18`

CLC	Clear Carry Flag	AT90S1200/2313/4414/8515

Syntax	CLC
Operands	none
Operation	C ← 0
Flags Affected	C
Encoding	`1001 0100 1000 1000`
Description	Clears the Carry flag (C) in SREG (status register).
Words	1
Cycles	1
Example	`clc  ; Clear carry flag`

CLH	Clear Half Carry Flag	AT90S1200/2313/4414/8515

Syntax	CLH
Operands	none
Operation	H ← 0
Flags Affected	H
Encoding	`1001 0100 1101 1000`
Description	Clears the Half Carry flag (H) in SREG (status register).
Words	1
Cycles	1
Example	`clh ; Clear the Half Carry flag`

CLI	*Clear Global Interrupt Flag*	*AT90S1200/2313/4414/8515*

Syntax	CLI
Operands	none
Operation	I ← 0
Flags Affected	I
Encoding	1001 0100 1111 1000
Description	Clears the Global Interrupt flag (I) in SREG (status register).
Words	1
Cycles	1
Example	cli ; Disable interrupts
	in r11,$16 ; Read port B
	sei ; Enable interrupts

CLN	*Clear Negative Flag*	*AT90S1200/2313/4414/8515*

Syntax	CLN
Operands	none
Operation	N ← 0
Flags Affected	N
Encoding	1001 0100 1010 1000
Description	Clears the Negative flag (N) in SREG (status register).
Words	1
Cycles	1
Example	cln ; Clear negative flag

CLR	*Clear Register*	*AT90S1200/2313/4414/8515*

Syntax	CLR Rd
Operands	0 ≤ d ≤ 31
Operation	Rd ← 0
Flags Affected	S, V, N, Z
Encoding	0010 01dd dddd dddd
Description	Clears a register. This instruction performs an exclusive OR between a register and itself. This will clear all bits in the register.
Words	1
Cycles	1
Example	clr r18 ; clear r18
	loop: inc r18 ; increase r18
	. . .
	cpi r18,$50 ; Compare r18 to $50
	brne loop

CLS	Clear Sign Flag	AT90S1200/2313/4414/8515

Syntax	CLS
Operands	none
Operation	S ← 0
Flags Affected	S
Encoding	1001 0100 1100 1000
Description	Clears the Sign flag (S) in SREG (status register).
Words	1
Cycles	1
Example	cls ; Clear sign flag

CLT	Clear T Flag	AT90S1200/2313/4414/8515

Syntax	CLT
Operands	none
Operation	T ← 0
Flags Affected	T
Encoding	1001 0100 1110 1000
Description	Clears the T flag in SREG (status register).
Words	1
Cycles	1
Example	clt ; Clear T flag

CLV	Clear Overflow Flag	AT90S1200/2313/4414/8515

Syntax	CLV
Operands	none
Operation	V ← 0
Flags Affected	V
Encoding	1001 0100 1011 1000
Description	Clears the Overflow flag (V) in SREG (status register).
Words	1
Cycles	1
Example	clv ; Clear overflow flag

CLZ	Clear Zero Flag	AT90S1200/2313/4414/8515

Syntax	CLZ
Operands	none
Operation	Z ← 0
Flags Affected	Z
Encoding	1001 0100 1111 1000
Description	Clears the Zero flag (Z) in SREG (status register).
Words	1
Cycles	1
Example	clz ; Clear zero

COM	One's Complement	AT90S1200/2313/4414/8515

Syntax	COM Rd
Operands	0 ≤ d ≤ 31
Operation	Rd ← $FF—Rd
Flags Affected	S, V, N, Z, C
Encoding	1001 010d dddd 0000
Description	Performs a one's-complement of register Rd.
Words	1
Cycles	1
Example	com r4 ; Take one's-complement of r4

CP	Compare	AT90S1200/2313/4414/8515

Syntax	CP Rd, Rr
Operands	0 ≤ d ≤ 31, 0 ≤ r ≤ 31
Operation	Rd-Rr
Flags Affected	H, S, V, N, Z, C
Encoding	0001 01rd dddd rrrr
Description	Performs a compare between two registers Rd and Rr. None of the registers are changed. All conditional branches can be used after this instruction.
Words	1
Cycles	1
Example	cp r4,r19 ; Compare r4 with r19

CPC	Compare with Carry	AT90S1200/2313/4414/8515

Syntax	CPC Rd, Rr
Operands	0 ≤ d ≤ 31, 0 ≤ r ≤ 31
Operation	Rd-Rr-C
Flags Affected	H, S, V, N, Z, C
Encoding	0000 01rd dddd rrrr
Description	Performs a compare between two registers Rd and Rr and also takes into account the previous carry. None of the registers are changed. All conditional branches can be used after this instruction.
Words	1
Cycles	1
Example	; Compare r3:r2 with r1:r0
	cp r2,r0 ; Compare low byte
	cpc r3,r1 ; Compare high byte

CPI	Compare with Immediate	AT90S1200/2313/4414/8515

Syntax	CPI Rd, K
Operands	16 ≤ d ≤ 31, 0≤ K ≤ 255
Operation	Rd-K

CPI	*Compare with Immediate*	*AT90S1200/2313/4414/8515*

Flags Affected	H, S, V, N, Z, C
Encoding	`0011 KKKK dddd KKKK`
Description	Performs a compare between register Rd and a constant. The register is not changed. All conditional branches can be used after this instruction.
Words	1
Cycles	1
Example	`cpi r19,3 ; Compare r19 with 3`

CPSE	*Compare Skip If Equal*	*AT90S1200/2313/4414/8515*

Syntax	`CPSE Rd, Rr`
Operands	$0 \le d \le 31$, $0 \le r \le 31$
Operation	Skips the next instruction if Rd ← Rr
Flags Affected	none
Encoding	`0001 00rd dddd rrrr`
Description	Performs a compare between two registers Rd and Rr, and skips the next instruction if Rd ← Rr.
Words	1
Cycles	1
Example	`cpse r4,r0 ; Compare r4 to r0`
	`neg r4    ; Only executed if r4<>r0`
	`nop       ; Continue (do nothing)`

DEC	*Decrement*	*AT90S1200/2313/4414/8515*

Syntax	`DEC Rd`
Operands	$0 \le d \le 31$
Operation	Rd ← Rd-1
Flags Affected	S, V, N, Z
Encoding	`1001 010d dddd 1010`
Description	Subtracts 1 from the contents of register Rd and places the result in the destination register Rd.
	The C flag in SREG is not affected by the operation, thus allowing the DEC instruction to be used on a loop counter in multiple-precision computations.
	When operating on unsigned values, only BREQ and BRNE branches can be expected to perform consistently. When operating on two's-complement values, all signed branches are available.
Words	1
Cycles	1
Example	`         ldi r17,$10 ; Load constant in r17`
	`loop: add r1,r2   ; Add r2 to r1`
	`      dec r17     ; Decrement r17`
	`      brne loop   ; Branch if r17<>0`
	`      nop         ; Continue (do nothing)`

AVR RISC Microcontroller Handbook

EOR	Exclusive OR	AT90S1200/2313/4414/8515

Syntax	EOR Rd, Rr
Operands	$0 \leq d \leq 31, 0 \leq r \leq 31$
Operation	Rd ← Rd ⊕ Rr
Flags Affected	S, V, N, Z
Encoding	0010 01rd dddd rrrr
Description	Performs logical EOR between the contents of register Rd and register Rr and places the result in the destination register Rd.
Words	1
Cycles	1
Example	eor r4,r4 ; Clear r4
	eor r0,r22 ; Bitwise exclusive OR between r0 and r22

ICALL	Indirect Call to Subroutine	AT90S1200/2313/4414/8515

Syntax	ICALL
Operands	none
Operation	PC(15-0) ← Z(15-0)
Flags Affected	none
Encoding	1001 0101 XXXX 1001
Description	Indirect call of a subroutine pointed to by the Z (16 bits) pointer register in the register file. The Z pointer register is 16 bits wide and allows call to a subroutine within the current 64K words (128K bytes) section in the program memory space.
Words	1
Cycles	3
Example	mov r30,r0 ; Set offset to call table
	icall ; Call routine pointed to by r31:r30

IJMP	Indirect Jump	AT90S1200/2313/4414/8515

Syntax	ICALL
Operands	none
Operation	PC(15-0) ← Z(15-0)
Flags Affected	none
Encoding	1001 0100 XXXX 1001
Description	Indirect jump to the address pointed to by the Z (16 bits) pointer register in the register file. The Z pointer register is 16 bits wide and allows jump within the current 64K words (128K bytes) section of program memory.
Words	1
Cycles	2
Example	mov r30,r0 ; Set offset to jump table
	ijmp ; Jump to routine pointed to by r31:r30

IN	Load an I/O Port to Register	AT90S1200/2313/4414/8515

Syntax	`IN Rd, P`
Operands	$0 \le d \le 31, 0 \le P \le 63$
Operation	$Rd \leftarrow P$
Flags Affected	none
Encoding	`1011 0PPd dddd PPPP`
Description	Loads data from the I/O Space (ports, timers, configuration registers, etc.) into register Rd in the register file.
Words	1
Cycles	1
Example	`in r25,$16 ; Read Port B`

INC	Increment	AT90S1200/2313/4414/8515

Syntax	`INC Rd`
Operands	$0 \le d \le 31$
Operation	$Rd \leftarrow Rd + 1$
Flags Affected	S, V, N, Z
Encoding	`1001 010d dddd 1011`
Description	Adds 1 to the contents of register Rd and places the result in the destination register Rd.
	The C flag in SREG is not affected by the operation, thus allowing the INC instruction to be used on a loop counter in multiple-precision computations.
	When operating on unsigned numbers, only BREQ and BRNE branches can be expected to perform consistently. When operating on two's-complement values, all signed branches are available.
Words	1
Cycles	1
Example	```

```
            clr r22      ; clear r22
      loop: inc r22      ; increment r22
            . . .
            cpi r22,$4F  ; Compare r22 to $4f
            brne loop    ; Branch if not equal
            nop          ; Continue (do nothing)
```

JMP	Jump	AT90S1200/2313/4414/8515

Syntax	`JMP k`
Operands	$0 \le k \le 4M$
Operation	$PC \leftarrow k$
Flags Affected	none
Encoding	`1001 010k kkkk 110k kkkk kkkk kkkk kkkk`
Description	Jump to an address within the entire 4M (words) program memory. See also RJMP.
Words	2

| Cycles | 3 |
| Example | |

```
                mov r1,r0  ; Copy r0 to r1
                jmp farplc ; Unconditional jump
                . . .
         farplc: nop        ; Jump destination (do nothing)
```

LD	Load Indirect from SRAM to Register Using Index X	AT90S1200/2313/4414/8515

Syntax	LD Rd, X X: Unchanged

```
         LD Rd, X     X: Unchanged
         LD Rd, X+    X: Post-incremented
         LD Rd,-X     X: Pre-incremented
```

Operands	$0 \le d \le 31$
Operation	X Unchanged Rd ← (X)
	X Post-incremented Rd ← (X) X ← X + 1
	X Pre-decremented X ← X -1 Rd ← (X)
Flags Affected	none
Encoding	X Unchanged 1001 000d dddd 1100
	X Post-incremented 1001 000d dddd 1101
	X Pre-decremented 1001 000d dddd 1110

Description   Loads 1 byte indirect from SRAM to register. The SRAM location is pointed to by the X (16 bits) pointer register in the register file. Memory access is limited to the current SRAM page of 64K bytes.

To access another SRAM page, the RAMPX in register in the I/O area has to be changed.

The X pointer register can be left unchanged after the operation, or it can be incremented or decremented. These features are especially suited for accessing arrays, tables, and stack pointer usage of the X pointer register.

Words	1
Cycles	2
Example	

```
         clr r27      ; Clear X high byte
         ldi r26,$20  ; Set X low byte to $20
         ld r0,X+     ; Load r0 with SRAM loc. $20(X post inc)
         ld r1,X      ; Load r1 with SRAM loc. $21
         ldi r26,$23  ; Set X low byte to $23
         ld r2,X      ; Load r2 with SRAM loc. $23
         ld r3,-X     ; Load r3 with SRAM loc. $22(X pre dec)
```

LD (LDD)	Load Indirect from SRAM to Register Using Index Y	AT90S1200/2313/4414/8515

Syntax	LD Rd, Y Y: Unchanged

```
         LD Rd, Y       Y: Unchanged
         LD Rd, Y+      Y: Post-incremented
         LD Rd,-Y       Y: Pre-incremented
         LDD Rd, Y+q    Y: Unchanged, q: Displacement
```

LD (LDD)	Load Indirect from SRAM to Register Using Index Y	AT90S1200/2313/4414/8515

Operands	$0 \le d \le 31, 0 \le q \le 63$	
Operation	Y Unchanged	Rd ← (Y)
	Y Post-incremented	Rd ← (Y) Y ← Y + 1
	Y Pre-decremented	Y ← Y-1 Rd ← (Y)
	Y Unchanged, q displacement	Rd ← (Y+q)
Flags Affected	none	
Encoding	Y Unchanged	1000 000d dddd 1000
	Y Post-incremented	1001 000d dddd 1001
	Y Pre-decremented	1001 000d dddd 1010
	Y Unchanged, q displacement	10q0 qq0d dddd 1qqq
Description	Loads 1 byte indirect with or without displacement from SRAM to register. The SRAM location is pointed to by the Y (16 bits) pointer register in the register file. Memory access is limited to the current SRAM page, of 64K bytes. To access another SRAM page the RAMPY register in the I/O area has to be changed.	
	The Y pointer register can be left unchanged after the operation, or it can be incremented or decremented. These features are especially suited for accessing arrays, tables, and stack pointer usage of the Y pointer register.	
Words	1	
Cycles	2	
Example		

```
clr r29      ; Clear Y high byte
ldi r28,$20  ; Set Y low byte to $20
ld r0,Y+     ; Load r0 with SRAM loc. $20(Y post inc)
ld r1,Y      ; Load r1 with SRAM loc. $21
ldi r28,$23  ; Set Y low byte to $23
ld r2,Y      ; Load r2 with SRAM loc. $23
ld r3,-Y     ; Load r3 with SRAM loc. $22(Y pre dec)
ldd r4,Y+2   ; Load r4 with SRAM loc. $24
```

LD (LDD)	Load Indirect from SRAM to Register Using Index Z	AT90S1200/2313/4414/8515

Syntax	LD Rd, Z	Z: Unchanged	
	LD Rd, Z+	Z: Post-incremented	(not S1200)
	LD Rd,-Z	Z: Pre-incremented	(not S1200)
	LDD Rd, Z+q	Z: Unchanged, q: Displacement	(not S1200)
Operands	$0 \le d \le 31, 0 \le q \le 63$		
Operation	Z Unchanged	Rd ← (Z)	
	Z Post-incremented	Rd ← (Z) Z ← Z + 1 (not S1200)	
	Z Pre-decremented	Z ← Z-1 Rd ← (Z) (not	

S1200)

	Z Unchanged, q displacement Rd ← (Z+q)	(not S1200)

Flags Affected	none	
Encoding	Z Unchanged	1000 000d dddd 0000
	Z Post-incremented	1001 000d dddd 0001 (not S1200)
	Z Pre-decremented	1001 000d dddd 0010 (not S1200)
	Z Unchanged, q displacement	10q0 qq0d dddd 0qqq (not S1200)

Description Loads 1 byte indirectly with or without displacement from SRAM to register. The SRAM location is pointed to by the Z (16 bits) pointer register in the register file. Memory access is limited to the current SRAM page of 64K bytes. To access another SRAM page the RAMPZ register in the I/O area has to be changed.

The Z pointer register can be left unchanged after the operation, or it can be incremented or decremented. These features are especially suited for stack pointer usage of the Z pointer register; however, because the Z pointer register can be used for indirect subroutine calls, indirect jumps, and table lookup, it is often more convenient to use the X or Y pointer as a dedicated stack pointer.

For using the Z pointer for table lookup in program memory, see the LPM instruction.

Words   1
Cycles   2
Example
```
clr r31     ; Clear Z high byte
ldi r30,$20 ; Set Z low byte to $20
ld r0,Z+    ; Load r0 with SRAM loc. $20 (Z post inc)
ld r1,Z     ; Load r1 with SRAM loc. $21
ldi r30,$23 ; Set Z low byte to $23
ld r2,Z     ; Load r2 with SRAM loc. $23
ld r3,-Z    ; Load r3 with SRAM loc. $22 (Z pre dec)
ldd r4,Z+2  ; Load r4 with SRAM loc. $24
```

LDI	*Load Immediate*	*AT90S1200/2313/4414/8515*

Syntax   LDI Rd, K
Operands $16 \leq d \leq 31$, $0 \leq K \leq 255$
Operation   Rd ← K
Flags Affected none
Encoding   1110 KKKK dddd KKKK
Description   Loads an 8-bit constant directly to registers 16 to 31.
Words   1
Cycles   1
Example
```
clr r31     ; Clear Z high byte
ldi r30,$F0 ; Set Z low byte to $F0
lpm         ; Load constant from program
            ; memory pointed to by Z
```

LDS	Load Direct from SRAM	AT90S1200/2313/4414/8515

Syntax	`LDS Rd, k`
Operands	$0 \le d \le 31, 0 \le k \le 65535$
Operation	Rd ← (k)
Flags Affected	none
Encoding	`1001 000d dddd 0000  kkkk kkkk kkkk kkkk`
Description	Loads 1 byte from the SRAM to a Register. A 16-bit address must be supplied.
	Memory access is limited to the current SRAM Page of 64K bytes. The LDS instruction uses the RAMPZ register to access memory above 64K bytes.
Words	2
Cycles	4
Example	`lds r2,$FF00 ; Load r2 with the contents of SRAM`
	`             ; location $FF00`
	`add r2,r1    ; add r1 to r2`
	`sts $FF00,r2 ; Write back`

LPM	Load Program Memory	AT90S1200/2313/4414/8515

Syntax	`LPM`
Operands	none
Operation	R0 ← (Z)
Flags Affected	none
Encoding	`1001 0101 110X 1000`
Description	Z points to program memory
Words	1
Cycles	3
Example	`clr r31     ; Clear Z high byte`
	`ldi r30,$F0 ; Set Z low byte`
	`lpm         ; Load constant from program`
	`            ; memory pointed to by Z (r31:r30)`

LSL	Logical Shift Left	AT90S1200/2313/4414/8515

Syntax	`LSL Rd`
Operands	$0 \le d \le 31$
Operation	

```
C    B7  B6  B5  B4  B3  B2  B1
```

Flags Affected	H, S, V, N, Z, C
Encoding	`0000 11dd dddd dddd`
Description	Shifts all bits in Rd one place to the left. Bit 0 is cleared. Bit 7 is loaded into the C flag of the SREG. This operation effectively multiplies an unsigned value by 2.

Words	1
Cycles	1
Example	lsl r0 ; Multiply r0 by 2

| **LSR** | *Logical Shift Right* | *AT90S1200/2313/4414/8515* |

Syntax	LSR Rd
Operands	$0 \le d \le 31$
Operation	

B7 B6 B5 B4 B3 B2 B1   C

Flags Affected	S, V, N, Z, C
Encoding	1001 010d dddd 0110
Description	Shifts all bits in Rd one place to the right. Bit 7 is cleared. Bit 0 is loaded into the C flag of the SREG. This operation effectively divides an unsigned value by 2. The C flag can be used to round the result.
Words	1
Cycles	1
Example	lsr r0 ; Divide r0 by 2

| **MOV** | *Copy Register* | *AT90S1200/2313/4414/8515* |

Syntax	MOV Rd,Rr
Operands	$0 \le d \le 31$, $0 \le r \le 31$
Operation	Rd ← Rr
Flags Affected	none
Encoding	0010 11rd dddd rrrr
Description	This instruction makes a copy of one register into another. The source register Rr is left unchanged; the destination register Rd is loaded with a copy of Rr.
Words	1
Cycles	1
Example	mov r16,r0 ; Copy r0 to r16

| **MUL** | *Multiply* | *AT90S1200/2313/4414/8515* |

Syntax	MUL Rd, Rr
Operands	$0 \le d \le 31$, $0 \le r \le 31$
Operation	R1,R0 ← Rr * Rd
Flags Affected	C
Encoding	1001 11rd dddd rrrr
Description	The multiplicand Rr and the multiplier Rd are two registers. The 16-bit product is placed in R1 (high byte) and R0 (low byte).

Handling the Hardware Resources

MUL	*Multiply*	*AT90S1200/2313/4414/8515*

Note that if the multiplicand and the multiplier are selected from R0 or R1, the result will overwrite those after multiplication.

Words	1
Cycles	2
Example	

```
mul r6,r5 ; Multiply r6 and r5
mov r6,r1 ; Copy result back in r6:r5
mov r5,r0 ; Copy result back in r6:r5
```

NEG	*Two's-Complement*	*AT90S1200/2313/4414/8515*

Syntax	NEG Rd
Operands	0 ≤ d ≤ 31
Operation	Rd ← $00-Rd
Flags Affected	H, S, V, N, Z, C
Encoding	1001 010d dddd 0001
Description	Replaces the contents of register Rd with its two's-complement; the value $80 is left unchanged.
Words	1
Cycles	1
Example	

```
          sub r11,r0   ; Subtract r0 from r11
          brpl positive ; Branch if result positive
          neg r11      ; Take two's-complement of r11
positive: nop          ; Branch destination
```

NOP	*No Operation*	*AT90S1200/2313/4414/8515*

Syntax	NOP
Operands	none
Operation	No
Flags Affected	none
Encoding	0000 0000 0000 0000
Description	Performs a single-cycle No Operation.
Words	1
Cycles	1
Example	

```
clr r16     ; Clear r16
ser r17     ; Set r17
out $18,r16 ; Write zeros to Port B
nop         ; Wait one cycle (do nothing)
out $18,r17 ; Write ones to Port B
```

OR	*Logical OR*	*AT90S1200/2313/4414/8515*

Syntax	OR Rd,Rr
Operands	0 ≤ d ≤ 31, 0 ≤ r ≤ 31
Operation	Rd ← Rd OR Rr
Flags Affected	S, V, N, Z
Encoding	0010 10rd dddd rrrr
Description	Performs the logical OR between the contents of register Rd and those of register Rr, and places the result in the destination register Rd.
Words	1
Cycles	1
Example	or r15,r16 ; Do bitwise OR between registers

ORI	*Logical OR with Immediate*	*AT90S1200/2313/4414/8515*

Syntax	ORI Rd, K
Operands	16 ≤ d ≤ 31, 0 ≤ K ≤ 255
Operation	Rd ← Rd OR K
Flags Affected	S, V, N, Z
Encoding	0110 KKKK dddd KKKK
Description	Performs the logical OR between the contents of register Rd and a constant, and places the result in the destination register Rd.
Words	1
Cycles	1
Example	ori r16,$F0 ; Set high nibble of r16 ori r17,1 ; Set bit 0 of r17

OUT	*Store Register to I/O Port*	*AT90S1200/2313/4414/8515*

Syntax	OUT P,Rr
Operands	0 ≤ r ≤ 31, 0 ≤ P ≤ 63
Operation	P ← Rr
Flags Affected	none
Encoding	1011 1PPr rrrr PPPP
Description	Stores data from register Rr in the register file to I/O space (ports, timers, configuration registers, etc.).
Words	1
Cycles	1
Example	clr r16 ; Clear r16 ser r17 ; Set r17 out $18,r16 ; Write zeros to Port B nop ; Wait (do nothing) out $18,r17 ; Write ones to Port B

POP

POP	*Pop Register from Stack*	*AT90S1200/2313/4414/8515*

Syntax ·   `POP rd`
Operands $0 \leq d \leq 31$
Operation   Rd ← STACK
Flags Affected   none
Encoding   `1001 000d dddd 1111`
Description   Loads register Rd with a byte from the STACK.
Words   1
Cycles   2
Example

```
              call routine ; Call subroutine
              ...
    routine: push r14     ; Save r14 on the stack
             push r13     ; Save r13 on the stack
             ...
             pop r13      ; Restore r13
             pop r14      ; Restore r14
             ret          ; Return from subroutine
```

PUSH

PUSH	*Push Register on Stack*	*AT90S1200/2313/4414/8515*

Syntax   `PUSH Rr`
Operands $0 \leq r \leq 31$
Operation   STACK ← Rr
Flags Affected   none
Encoding   `1001 001d dddd 1111`
Description   Stores the contents of register Rr on the STACK.
Words   1
Cycles   2
Example

```
              call routine ; Call subroutine
              ...
    routine: push r14     ; Save r14 on the stack
             push r13     ; Save r13 on the stack
             ...
             pop r13      ; Restore r13
             pop r14      ; Restore r14
             ret          ; Return from subroutine
```

RCALL

RCALL	*Relative Call to Subroutine*	*AT90S1200/2313/4414/8515*

Syntax   `RCALL k`
Operands $-2K \leq k \leq 2K$
Operation   PC ← PC + k + 1
Flags Affected   none
Encoding   `1101 kkkk kkkk kkkk`

Description	Calls a subroutine within ±2K words (4K bytes). The return address (the instruction after the RCALL) is stored onto the stack. (See also CALL.)
Words	1
Cycles	3
Example	

```
          rcall routine ; Call subroutine
          . . .
routine:  push r14      ; Save r14 on the stack
          . . .
          pop r14       ; Restore r14
          ret           ; Return from subroutine
```

RET	Return from Subroutine	AT90S1200/2313/4414/8515

Syntax	RET
Operands	none
Operation	PC ← STACK
Flags Affected	none
Encoding	1001 0101 0XX0 1000
Description	Returns from subroutine. The return address is loaded from the STACK.
Words	1
Cycles	4
Example	

```
          call routine ; Call subroutine
          . . .
routine:  push r14      ; Save r14 on the stack
          . . .
          pop r14       ; Restore r14
          ret           ; Return from subroutine
```

RETI	Return from Interrupt	AT90S1200/2313/4414/8515

Syntax	RETI
Operands	none
Operation	PC ← STACK
Flags Affected	I
Encoding	1001 0101 0XX1 1000
Description	Returns from interrupt. The return address is loaded from the STACK and the global interrupt flag is set.
Words	1
Cycles	4
Example	

```
          . . .
extint:   push r0 ; Save r0 on the stack
          . . .
          pop r0  ; Restore r0
          reti    ; Return and enable interrupts
```

Handling the Hardware Resources   99

RJMP	Relative Jump	AT90S1200/2313/4414/8515

Syntax	`RJMP k`
Operands	$-2K \le k \le 2K$
Operation	$PC \leftarrow PC + k + 1$
Flags Affected	none
Encoding	`1100 kkkk kkkk kkkk`
Description	Relative jump to an address within PC – 2K and PC + 2K (words). In the assembler, labels are used instead of relative operands. For AVR microcontrollers with program memory not exceeding 4K words (8K bytes) this instruction can address the entire memory from every address location.
Words	1
Cycles	2
Example	

```
        cpi r16,$42  ; Compare r16 to $42
        brne error   ; Branch if r16 <> $42
        rjmp ok      ; Unconditional branch
error:  add r16,r17  ; Add r17 to r16
        inc r16      ; Increment r16
ok:     nop          ; Destination for rjmp (do nothing)
```

ROL	Rotate Left through Carry	AT90S1200/2313/4414/8515

Syntax	`ROL Rd`
Operands	$0 \le d \le 31$
Operation	

C   B7 B6 B5 B4 B3 B2 B1   C

Flags Affected	H, S, V, N, Z, C
Encoding	`0001 11dd dddd dddd`
Description	Shifts all bits in Rd one place to the left. The C flag is shifted into bit 0 of Rd. Bit 7 is shifted into the C flag.
Words	1
Cycles	1
Example	`rol r15 ; Rotate left`

ROR	Rotate Right through Carry	AT90S1200/2313/4414/8515

Syntax	`ROR Rd`
Operands	$0 \le d \le 31$
Operation	

C   B7 B6 B5 B4 B3 B2 B1   C

Flags Affected	S, V, N, Z, C
Encoding	`1001 010d dddd 01111`

Description	Shifts all bits in Rd one place to the right. The C flag is shifted into bit 7 of Rd. Bit 0 is shifted into the C flag.
Words	1
Cycles	1
Example	`ror r15 ; Rotate right`

SBC	*Subtract with Carry*	*AT90S1200/2313/4414/8515*

Syntax	`SBC Rd, Rr`
Operands	$0 \le d \le 31$, $0 \le r \le 31$
Operation	Rd ← Rd-Rr -C
Flags Affected	H, S, V, N, Z, C
Encoding	`0000 10rd dddd rrrr`
Description	Subtracts two registers and subtracts with the C flag and places the result in the destination register Rd.
Words	1
Cycles	1
| Example | ```
          ; Subtract r1:r0 from r3:r2
sub r2,r0 ; Subtract low byte
sbc r3,r1 ; Subtract with carry high byte
``` |

SBCI	*Subtract Immediate with Carry*	*AT90S1200/2313/4414/8515*

Syntax	`SBCI Rd, K`
Operands	$16 \le d \le 31$, $0 \le K \le 255$
Operation	Rd ← Rd-K-C
Flags Affected	H, S, V, N, Z, C
Encoding	`0100 KKKK dddd KKKK`
Description	Subtracts a constant from a register and subtracts with the C flag and places the result in the destination register Rd.
Words	1
Cycles	1
| Example | ```
            ; Subtract $4F23 from r17:r16
subi r16,$23 ; Subtract low byte
sbci r17,$4F ; Subtract with carry high byte
``` |

SBI	*Set Bit in I/O Register*	*AT90S1200/2313/4414/8515*

Syntax	`SBI P,b`
Operands	$0 \le P \le 31$, $0 \le b \le 7$
Operation	I/O(P,b) ← 1
Flags Affected	none
Encoding	`1001 1010 PPPP Pbbb`
Description	Sets a specified bit in an I/O register. This instruction operates on the lower 32 I/O registers—addresses 0-31.
Words	1

SBI	Set Bit in I/O Register	AT90S1200/2313/4414/8515

Cycles	2
| Example | ```
out $1E,r0 ; Write EEPROM address
sbi $1C,0  ; Set read bit in EECR
in r1,$1D  ; Read EEPROM data
``` |

SBIC	Skip If Bit in I/O Register Is Cleared	AT90S1200/2313/4414/8515

Syntax	`SBIC P,b`
Operands	$0 \le P \le 31$, $0 \le b \le 7$
Operation	Skip next instruction if I/O(P,b) = 0
Flags Affected	none
Encoding	`1001 1001 PPPP Pbbb`
Description	Tests a single bit in an I/O register and skips the next instruction if the bit is cleared. This instruction operates on the lower 32 I/O registers—addresses 0-31.
Words	1
Cycles	2 (false); 3 (true)
| Example | ```
e2wait: sbic $1C,1  ; Skip next inst. if EEWE cleared
        rjmp e2wait ; EEPROM write not finished
        nop         ; Continue (do nothing)
``` |

SBIS	Skip If Bit in I/O Register Is Set	AT90S1200/2313/4414/8515

Syntax	`SBIS P,b`
Operands	$0 \le P \le 31$, $0 \le b \le 7$
Operation	Skip next instruction if I/O(P,b) = 1
Flags Affected	none
Encoding	`1001 1011 PPPP Pbbb`
Description	Tests a single bit in an I/O register and skips the next instruction if the bit is set. This instruction operates on the lower 32 I/O registers—addresses 0-31.
Words	1
Cycles	2 (false); 3 (true)
| Example | ```
waitset: sbis $10,0  ; Skip next inst.
                     ;if bit 0 in Port D set
         rjmp waitset ; Bit not set
         nop          ; Continue (do nothing)
``` |

SBIW	Subtract Immediate from Word	AT90S1200/2313/4414/8515

Syntax	`SBIW Rdl, K`
Operands	dl $\in$ {24,26,28,30}, $0 \le K \le 63$
```

Operation	Rdh:Rdl ← Rdh:Rdl-K
Flags Affected	S, V, N, Z, C
Encoding	`1001 0111 KKdd KKKK`
Description	Subtracts an immediate value (0–63) from a register pair and places the result in the register pair. This instruction operates on the upper four register pairs and is well suited for operations on the pointer registers.
Words	1
Cycles	2
Example	`sbiw r24,1  ; Subtract 1 from r25:r24` `sbiw r28,63 ; Subtract 63 from the Y pointer(r29:r28)`

SBR	*Set Bits in Register*	*AT90S1200/2313/4414/8515*

Syntax	`SBR Rd, K`
Operands	16 ≤ d ≤ 31, 0 ≤ K ≤ 255
Operation	Rd ← Rd OR K
Flags Affected	S, V, N, Z
Encoding	`0110 KKKK dddd KKKK`
Description	Sets specified bits in register Rd. Performs the logical ORI between the contents of register Rd and a constant mask K and places the result in the destination register Rd.
Words	1
Cycles	1
Example	`sbr r16,3   ; Set bits 0 and 1 in r16` `sbr r17,$F0 ; Set 4 MSB in r17`

SBRC	*Skip If Bit in* *Register Is Cleared*	*AT90S1200/2313/4414/8515*

Syntax	`SBRC Rr,b`
Operands	0 ≤ r ≤ 31, 0 ≤ b ≤ 7
Operation	Skip next instruction if Rr(b) = 0
Flags Affected	none
Encoding	`1111 110r rrrr Xbbb`
Description	Tests a single bit in a register and skips the next instruction if the bit is cleared.
Words	1
Cycles	1 (false); 2 (true)

Example	`sub r0,r1 ; Subtract r1 from r0`
	`sbrc r0,7 ; Skip if bit 7 in r0 cleared`
	`sub r0,r1 ; Only executed if bit 7 in r0 not cleared`
	`nop       ; Continue (do nothing)`

SBRS	Skip If Bit in	AT90S1200/2313/4414/8515
	Register Is Set	

Syntax	SBRS Rr,b
Operands	$0 \leq r \leq 31, 0 \leq b \leq 7$
Operation	Skip next instruction if Rr(b) = 1
Flags Affected	none
Encoding	1111 111r rrrr Xbbb
Description	Tests a single bit in a register and skips the next instruction if the bit is set.
Words	1
Cycles	1 (false); 2 (true)
Example	sub r0,r1 ; Subtract r1 from r0
	sbrs r0,7 ; Skip if bit 7 in r0 set
	neg r0 ; Only executed if bit 7 in r0 not set
	nop ; Continue (do nothing)

SEC	Set Carry Flag	AT90S1200/2313/4414/8515

Syntax	SEC
Operands	none
Operation	C ← 1
Flags Affected	C
Encoding	1001 0100 0000 1000
Description	Sets the Carry flag (C) in SREG (status register).
Words	1
Cycles	1
Example	sec ; Set carry flag
	adc r0,r1 ; r0=r0+r1+1

SEH	Set Half Carry Flag	AT90S1200/2313/4414/8515

Syntax	SEH
Operands	none
Operation	H ← 1
Flags Affected	H
Encoding	1001 0100 0101 1000
Description	Sets the Half Carry flag (H) in SREG (status register).
Words	1
Cycles	1
Example	seh ; Set half carry flag

SEI	*Set Global Interrupt Flag*	*AT90S1200/2313/4414/8515*

Syntax	SEI
Operands	none
Operation	I ← 1
Flags Affected	I
Encoding	1001 0100 0111 1000
Description	Sets the Global Interrupt flag (I) in SREG (status register).
Words	1
Cycles	1
Example	cli ; Disable interrupts
	in r13,$16 ; Read Port B
	sei ; Enable interrupts

SEN	*Set Negative Flag*	*AT90S1200/2313/4414/8515*

Syntax	SEN
Operands	none
Operation	N ← 1
Flags Affected	N
Encoding	1001 0100 0110 1000
Description	Sets the Negative flag (N) in SREG (status register).
Words	1
Cycles	1
Example	sen ; Set negative flag

SER	*Set All Bits in Register*	*AT90S1200/2313/4414/8515*

Syntax	SER Rd
Operands	$16 \leq d \leq 31$
Operation	Rd ← $FF
Flags Affected	none
Encoding	1110 1111 dddd 1111
Description	Loads $FF directly to register Rd.
Words	1
Cycles	1
Example	clr r16 ; Clear r16
	ser r17 ; Set r17
	out $18,r16 ; Write zeros to Port B
	nop ; Delay (do nothing)
	out $18,r17 ; Write ones to Port B

SES	Set Signed Flag	*AT90S1200/2313/4414/8515*

Syntax	SES
Operands	none
Operation	S ← 1
Flags Affected	S
Encoding	1001 0100 0100 1000
Description	Sets the Signed flag (S) in SREG (status register).
Words	1
Cycles	1
Example	ses ; Set signed flag

SET	Set T Flag	*AT90S1200/2313/4414/8515*

Syntax	SET
Operands	none
Operation	T ← 1
Flags Affected	T
Encoding	1001 0100 0110 1000
Description	Sets the T flag in SREG (status register).
Words	1
Cycles	1
Example	set ; Set T flag

SEV	Set Overflow Flag	*AT90S1200/2313/4414/8515*

Syntax	SEV
Operands	none
Operation	V ← 1
Flags Affected	V
Encoding	1001 0100 0011 1000
Description	Sets the Overflow flag (V) in SREG (status register).
Words	1
Cycles	1
Example	sev ; Set overflow flag

SEZ	Set Zero Flag	*AT90S1200/2313/4414/8515*

Syntax	SEZ
Operands	none
Operation	Z ← 1
Flags Affected	Z
Encoding	1001 0100 0001 1000

Description	Sets the Zero flag (Z) in SREG (status register).
Words	1
Cycles	1
Example	`sez ; Set zero flag`

SLEEP	Sleep	AT90S1200/2313/4414/8515

Syntax	SLEEP
Operands	none
Operation	—
Flags Affected	none
Encoding	`1001 0100 0001 1000`
Description	Sets the circuit in sleep mode defined by the MCU control register. When an interrupt wakes up the MCU from a sleep state, the instruction following the SLEEP instruction will be executed before the interrupt handler is executed.
Words	1
Cycles	1
Example	`sleep ; Put MCU in sleep mode`

ST	Store Indirect from Register to SRAM Using Index X	AT90S1200/2313/4414/8515

Syntax	ST X, Rr	X: Unchanged
	ST X+, Rr	X: Post-incremented
	ST -X, Rr	X: Pre-decremented
Operands	$0 \leq r \leq 31$	
Operation	X Unchanged	$(X) \leftarrow Rr$
	X Post-incremented	$(X) \leftarrow Rr \quad X \leftarrow X + 1$
	X Pre-decremented	$X \leftarrow X—1 \quad (X) \leftarrow Rr$
Flags Affected	none	
Encoding	X Unchanged	`1001 001r rrrr 1100`
	X Post-incremented	`1001 001r rrrr 1101`
	X Pre-decremented	`1001 001r rrrr 1110`
Description	Stores one byte indirect from Register to SRAM. The SRAM location is pointed to by the X (16 bits) pointer register in the register file. Memory access is limited to the current SRAM page of 64K bytes. To access another SRAM page the RAMPX register in the I/O area has to be changed.	
	The X pointer register can be left unchanged after the operation, or it can be incremented or decremented. These features are especially suited for stack pointer usage of the X pointer register.	
Words	1	
Cycles	2	

	Store Indirect from Register to SRAM	
ST	Using Index X	AT90S2313/4414/8515

Example	```clr r27```	; Clear X high byte
	```ldi r26,$20```	; Set X low byte to $20
	```st X+,r0```	; Store r0 in SRAM loc. $20(X post inc)
	```st X,r1```	; Store r1 in SRAM loc. $21
	```ldi r26,$23```	; Set X low byte to $23
	```st r2,X```	; Store r2 in SRAM loc. $23
	```st r3,-X```	; Store r3 in SRAM loc. $22(X pre dec)

	Store Indirect from Register to SRAM	
ST (STD)	Using Index Y	AT90S1200/2313/4414/8515

Syntax	ST Y, Rr	Y: Unchanged
	ST Y+, Rr	Y: Post-incremented
	ST -Y, Rr	Y: Pre-incremented
	STD Y+q, Rr	Y: Unchanged, q: Displacement
Operands	$0 \leq r \leq 31, 0 \leq q \leq 63$	
Operation	Y Unchanged	$(Y) \leftarrow Rr$
	Y Post-incremented	$(Y) \leftarrow Rr \quad Y \leftarrow Y + 1$
	Y Pre-decremented	$Y \leftarrow Y\text{-}1 \quad (Y) \leftarrow Rr$
	Y Unchanged, q displacement	$(Y+q) \leftarrow Rr$
Flags Affected	none	
Encoding	Y Unchanged	1000 001r rrrr 1000
	Y Post-incremented	1001 001r rrrr 1001
	Y Pre-decremented	1001 001r rrrr 1010
	Y Unchanged, q displacement	10q0 qq1r rrrr 1qqq

Description: Stores one byte indirect with or without displacement from Register to SRAM. The SRAM location is pointed to by the Y (16 bits) pointer register in the register file. Memory access is limited to the current SRAM page of 64K bytes. To access another SRAM page, the RAMPY register in the I/O area has to be changed.

The Y pointer register can be left unchanged after the operation, or it can be incremented or decremented. These features are especially suited for stack pointer usage of the Y pointer register.

Words	1	
Cycles	2	
Example	```clr r29```	; Clear Y high byte
	```ldi r28,$20```	; Set Y low byte to $20
	```st Y+,r0```	; Store r0 in SRAM loc. $20(Y post inc)
	```st Y,r1```	; Store r1 in SRAM loc. $21
	```ldi r28,$23```	; Set Y low byte to $23
	```st Y,r2```	; Store r2 in SRAM loc. $23
	```st -Y,r3```	; Store r3 in SRAM loc. $22(Y pre dec)
	```std Y+2,r4```	; Store r4 in SRAM loc. $24

ST (STD)	Store Indirect from Register to SRAM Using Index Z		AT90S1200/2313/4414/8515

Syntax	`ST Z, Rr`	Z: Unchanged		
	`ST Z+, Rr`	Z: Post-incremented	(not S1200)	
	`ST -Z, Rr`	Z: Pre-incremented	(not S1200)	
	`STD Z+q, Rr`	Z: Unchanged, q: Displacement	(not S1200)	
Operands	$0 \le r \le 31, 0 \le q \le 63$			
Operation	Z Unchanged	(Z) ← Rr		
	Z Post-incremented	(Z) ← Rr	Z ← Z + 1 (not S1200)	
	Z Pre-decremented	Z ← Z—1	(Z) ← Rr (not S1200)	
	Z Unchanged, q displacement	(Z+q) ← Rr	(not S1200)	
Flags Affected	none			
Encoding	Z Unchanged	1000 001r rrrr 0000		
	Z Post-incremented	1001 001r rrrr 0001 (not S1200)		
	Z Pre-decremented	1001 001r rrrr 0010 (not S1200)		
	Z Unchanged, q displacement	10q0 qq1r rrrr 0qqq (not S1200)		

Description: Stores one byte indirect with or without displacement from Register to SRAM. The SRAM location is pointed to by the Z (16 bits) pointer register in the register file. Memory access is limited to the current SRAM page of 64K bytes. To access another SRAM page, the RAMPZ register in the I/O area has to be changed.

The Z pointer register can be left unchanged after the operation, or it can be incremented or decremented. These features are very well suited for stack pointer usage of the Z pointer register, but because the Z pointer register can be used for indirect subroutine calls, indirect jumps and table lookup, it is often more convenient to use the X or Y pointer as a dedicated stack pointer.

Words: 1

Cycles: 2

Example:
```
clr r31     ; Clear Z high byte
ldi r30,$20 ; Set Z low byte to $20
st Z+,r0    ; Store r0 in SRAM loc. $20(Z post inc)
st Z,r1     ; Store r1 in SRAM loc. $21
ldi r30,$23 ; Set Z low byte to $23
st Z,r2     ; Store r2 in SRAM loc. $23
st -Z,r3    ; Store r3 in SRAM loc. $22(Z pre dec)
std Z+2,r4  ; Store r4 in SRAM loc. $24
```

STS	Store Direct to SRAM	AT90S1200/2313/4414/8515

Syntax	`STS k, Rr`
Operands	$0 \le r \le 31, 0 \le k \le 65535$
Operation	(k) ← Rr
Flags Affected	none
Encoding	1001 001r rrrr 0000 kkkk kkkk kkkk kkkk

STS	Store Direct to SRAM	AT90S2313/4414/8515

Description	Stores one byte from a register to the SRAM. A 16-bit address must be supplied. Memory access is limited to the current SRAM page of 64K bytes.
	The SDS instruction uses the RAMPZ register to access memory above 64K bytes.
Words	2
Cycles	3
Example	

```
lds r2,$FF00 ; Load r2 with the contents of SRAM
             ; location $FF00
add r2,r1    ; add r1 to r2
sts $FF00,r2 ; Write back
```

SUB	Subtract without Carry	AT90S1200/2313/4414/8515

Syntax	SUB Rd, Rr
Operands	$0 \le d \le 31$, $0 \le r \le 31$
Operation	Rd ← Rd-Rr
Flags Affected	H, S, V, N, Z, C
Encoding	0001 10rd dddd rrrr
Description	Subtracts two registers and places the result in the destination register Rd.
Words	1
Cycles	1
Example	sub r13,r12 ; Subtract r12 from r13

SUBI	Subtract Immediate	AT90S1200/2313/4414/8515

Syntax	SUBI Rd, K
Operands	$16 \le d \le 31$, $0 \le K \le 255$
Operation	Rd ← Rd-K
Flags Affected	H, S, V, N, Z, C
Encoding	0101 KKKK dddd KKKK
Description	Subtracts a register and a constant and places the result in the destination register Rd. This instruction is working on registers R16 to R31 and is very well suited for operations on the X, Y, and Z pointers.
Words	1
Cycles	1
Example	subi r22,$11 ; Subtract $11 from r22

SWAP	Swap Nibbles	AT90S1200/2313/4414/8515

Syntax	SWAP Rd
Operands	0 ≤ d ≤ 31
Operation	R(7-4) ← Rd(3-0), R(3-0) ← Rd(7-4)
Flags Affected	none
Encoding	1001 010d dddd 0010
Description	Swaps high and low nibbles in a register.
Words	1
Cycles	1
Example	swap r1 ; Swap high and low nibble of r1

TST	Test for Zero or Minus	AT90S1200/2313/4414/8515

Syntax	TST Rd
Operands	0 ≤ d ≤ 31
Operation	Rd ← Rd AND Rd
Flags Affected	S, V, N, Z,
Encoding	0010 00dd dddd dddd
Description	Tests if a register is zero or negative. Performs a logical AND between a register and itself. The register will remain unchanged.
Words	1
Cycles	1
Example	tst r0 ; Test r0

WDR	Watchdog Reset	AT90S1200/2313/4414/8515

Syntax	WDR
Operands	none
Operation	Watchdog Timer Restart
Flags Affected	none
Encoding	1001 0101 101X 1000
Description	Resets the watchdog timer. This instruction must be executed within a limited time given by the WD prescaler. See the Watchdog Timer hardware specification.
Words	1
Cycles	1
Example	wdr ; Reset watchdog timer

3.3 Reset and Interrupt Handling

During reset, all I/O registers are set to their initial values, and the program starts execution from address $000. The instruction placed in address $000 must be an rjmp instruction to the reset handling routine placed anywhere in the program body. The string <instr> is equivalent to any instructions in the source.

```
.include "1200def.inc"
.device at90s1200
.cseg                        ; Code Segment begins at $000

        rjmp   RESET         ; Reset Handle

MAIN:   <instr>              ; Main program start
        <instr>
forever:rjmp   forever       ; Loop forever

RESET:  <instr>              ; Reset Handle
        rjmp   MAIN
```

This short example of source text shows the relative jump to the reset handle located anywhere in the program body—here, after the main part of the program. If the program never enables an interrupt source, the interrupt vectors are not used, and regular program code can be placed at these locations starting at address $0001.

If you like to prepare the source code for further usage of interrupts, then including an interrupt vector table from point of start can be helpful.

For AT90S1200, this vector table looks as follows:

```
$000   rjmp RESET           ; Reset Handle
$001   rjmp EXT_INT0        ; IRQ0 Handle
$002   rjmp TIM0_OVF        ; Timer0 overflow Handle
$003   rjmp ANA_COMP        ; Analog Comparator Handle
```

At address $000, the reset vector is placed again followed from the three possible interrupts for that microcontroller.

If you work only with the Timer0 Overflow interrupt, for example, then you have several possibilities:

1. Notify an interrupt handler for the unused interrupt containing only a reti instruction.

2. Replace the `rjmp` instruction of an unused interrupt by a `reti` instruction in the interrupt vector table.
3. Use the `.org` directive to place the interrupt call directly.

The changed interrupt vector table for the first two options look as follows:

```
.include "1200def.inc"
.device at90s1200

        rjmp    RESET           ; Reset Handle
        reti                    ; IRQ0 Handle
        rjmp    TIM0_OVF        ; Timer0 overflow Handle
        rjmp    ANA_COMP        ; Analog Comparator Handle

ANA_COMP:                       ; Analog Comparator Handle
        reti                    ; without any activity
TIM0_OVF:
        out     TCNT0, reload   ; Reload Timer/Counter0
        reti

RESET:<instr>                   ; Reset Handle

MAIN: <instr>                   ; Main program start
```

In our example, only the Timer/Counter0 Overflow Interrupt is used. Its handle reloads the Timer/Counter0 after an interrupt occurs.

The first option was used for the Analog Comparator Interrupt. The handle is notified but contains only the `reti` instruction. The second option was used for the External Interrupt. The `rjmp` instruction in the interrupt vector table was immediately replaced by a `reti` instruction.

The third option (using the .ORG directive) is explained with the next source. Before the relative jump to the Timer/Counter0 Overflow Interrupt handle is assembled, the memory location is adjusted by the `.org` directive. The use of the predefined symbolic address or the absolute address is an issue of taste.

```
.include "1200def.inc"
.device at90s1200

        rjmp    RESET           ; Reset Handle
.org OVF0addr                   ; Predefined in include file
.org 2                          ; the same as above
```

```
        rjmp    TIM0_OVF        ; Timer0 overflow Handle
TIM0_OVF:
        out     TCNT0, reload   ; Reload Timer/Counter0
        reti

RESET:<instr>                   ; Reset Handle

MAIN: <instr>                   ; Main program start
```

A more general program setup for the reset and interrupt vector addresses for the microcontrollers AT90S2313 is shown in the following listing:

```
$000    rjmp RESET          ; Reset Handle
$001    rjmp EXT_INT0       ; IRQ0 Handle
$002    rjmp EXT_INT1       ; IRQ1 Handle
$003    rjmp TIM1_CAPT      ; Timer1 capture Handle
$004    rjmp TIM1_COMP      ; Timer1 compare Handle
$005    rjmp TIM1_OVF       ; Timer1 overflow Handle
$006    rjmp TIM0_OVF       ; Timer0 overflow Handle
$007    rjmp UART_RXC       ; UART RX Complete Handle
$008    rjmp UART_DRE       ; UART UDR Empty Handle
$009    rjmp UART_TXC       ; UART TX Complete Handle
$00a    rjmp ANA_COMP       ; Analog Comparator Handle
                            ;
$00b  MAIN: <instr>         ; Main program start ... ... ...
```

The microcontrollers AT90S4414 or AT90S8515 have full interrupt capabilities. Therefore, their interrupt vector tables grow further:

```
$000    rjmp RESET          ; Reset Handle
$001    rjmp EXT_INT0       ; IRQ0 Handle
$002    rjmp EXT_INT1       ; IRQ1 Handle
$003    rjmp TIM1_CAPT      ; Timer1 capture Handle
$004    rjmp TIM1_COMPA ;   ; Timer1 compareA Handle
$005    rjmp TIM1_COMPB ;   ; Timer1 compareB Handle
$006    rjmp TIM1_OVF       ; Timer1 overflow Handle
$007    rjmp TIM0_OVF       ; Timer0 overflow Handle
$008    rjmp SPI_HANDLE ;   ; SPI TX Handle
$009    rjmp UART_RXC       ; UART RX Complete Handle
$00a    rjmp UART_DRE       ; UART UDR Empty Handle
$00b    rjmp UART_TXC       ; UART TX Complete Handle
$00c    rjmp ANA_COMP       ; Analog Comparator Handle
                            ;
$00d  MAIN: <instr>         ; Main program start ... ... ...
```

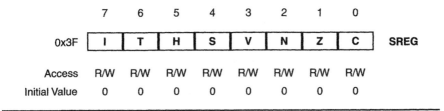

Figure 3-13
Status register SREG.

The flexible interrupt module has its control registers in the I/O space with an additional global interrupt enable bit (I) in the Status register. Figure 3-13 shows the Interrupt Enable bit I in the MSB position in the Status register (SREG).

The global interrupt enable bit must be set to enable any interrupt. The individual interrupt enable control is then performed in the interrupt mask registers—GIMSK and TIMSK.

If the global interrupt enable register is cleared, none of the interrupts are enabled independent of the GIMSK and TIMSK values.

The global interrupt enable bit is cleared by hardware after an interrupt has occurred and is set by the `reti` instruction to enable subsequent interrupts.

If an interrupt occurs, the return address program counter (PC) is stored on the stack described in the next section.

3.4 Watchdog Handling

The `wdr` instruction resets the watchdog timer. If the watchdog is enabled, this instruction must be executed within a limited time given by the watchdog prescaler; otherwise, the system will reset.

Optimum placement of watchdog refreshes within the program is no simple task and generally is the last procedure before the program is finalized. Normally, the user should examine the software flow and timing for all interrupt handlers and subroutines, critical and noncritical applications.

Ideally, one watchdog refresh in the whole program would be the best, but because microcontrollers have large programs and more on-chip functionality, this is rarely achievable. If possible, the watchdog refresh routines should never be placed in interrupt handlers or subroutines; they should be placed strategically in the main loop. Take care that the refresh rate is not too large or the chances of recovery from a runaway condition will be decreased.

3.5 Stack

The stack is a memory area working as last-in first-out memory for temporary data storage. During straightforward execution flow, the program counter is incremented after each instruction, fetching the next instruction from program memory.

Operating subroutine calls and interrupts means changing the order of operation for this task. For the operation to continue after the subroutine or interrupt is handled, that origin memory location or program counter contents must be stored during this time.

The AT90S1200 holds a three-level-deep hardware stack dedicated for subroutines and interrupts. The remaining microcontrollers of the AVR family allocate the stack in the general data SRAM, and consequently the stack size is only limited by the total SRAM size and the usage of the SRAM. For these microcontrollers, all user programs must initialize the stack pointer SP in the reset routine (before subroutines or interrupts are executed). The 16-bit stack pointer SP is read/write accessible in the I/O space (Figure 3-14).

The following program lines show the required stack pointer initialization as a part of the general initialization following reset.

```
RESET:   ldi    temp,   LOW(RAMEND)
         out    SPL,    temp          ; Initialize SPL
         ldi    temp,   HIGH(RAMEND)
         out    SPH,    temp          ; Initialize SPH
```

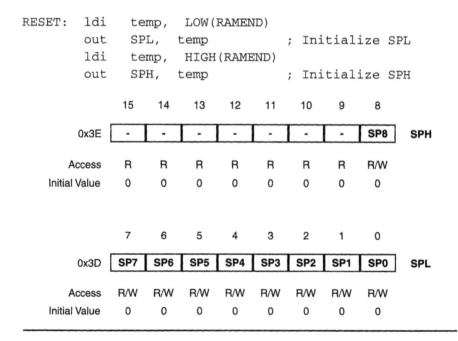

Figure 3-14
16-bit stack pointer, SPH and SPL.

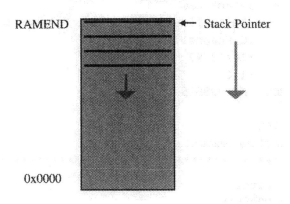

DATA MEMORY

Figure 3-15
Stack allocation in data memory.

RAMEND is defined in an include file (8515def.inc, for example) normally containing many directives; it holds the size of the SRAM available for data storage and the stack besides.

The AT90S8515 microcontroller has an SRAM capacity of 512 bytes, so it is addressed by 9 bits. Loading the SPL register only is not enough for this type of microcontroller. The SRAM capacity of the AT90S2313 and AT90S4414 is smaller, so the SPL register will do the initialization alone.

As Figure 3-15 shows, an allocation of the stack on the upper area of data memory is recommended because it will grow to lower addresses.

The stack can be used for temporary storage of data as well. In assembler programs, the declared variables are normally stored in registers. However, the number of available registers is limited. An apparent enhancement of the number of registers can be achieved by temporary storage of their unused contents on the stack. A simple example shows the mechanism.

On condition that all registers are used in the main program we have to do some data manipulations in an interrupt handler or a subroutine and need one register for that. To get a "free" register, during this data manipulation we save the contents of one unused register on the stack until the manipulation is finished. After this operation, the original conditions can be restored. Listing 3-1 shows a short program example written for simulation in AVR Studio.

```asm
;*********************************************************
;* File Name   :stack.asm
;* Title       :Initialization of Stack Pointer & Push & Pop
;* Author      :C.Kuehnel
;* Date        :10/18/97
;* Version     :1.0
;* Target MCU  :AT90S8515
;*
;* DESCRIPTION
;* Stack handling example
;*********************************************************

;***** Directives
.device    at90s8515
.nolist
.include   "8515def.inc"
.list

; Main Program Register variables
;-------------------------------------------------------
.def       temp            = r16

;***** Interrupt vector table
           rjmp   RESET          ; Reset handle

;***** Subroutines

manipulate_temp:
           push   temp            ; Save temp on stack
           ldi    temp, $A5       ; Manipulate the (local) temp
           swap   temp
           pop    temp            ; Restore the old temp
           ret

;***** Main
RESET: ldi       temp, LOW(RAMEND)
       out        SPL, temp       ; Initialize SPL
       ldi        temp, HIGH(RAMEND)
       out        SPH, temp       ; Initialize SPH

       ldi        temp, $FF       ; Load register temp
loop:  rcall      manipulate_temp

       rjmp       loop            ; Repeat forever
```

Listing 3-1
Initialization of stack pointer and push and pop (stack.asm).

After the label RESET, the stack pointer initialization follows as described earlier in this chapter. Figure 3-16 shows a screenshot of the AVR Studio after stack pointer initialization.

As shown in Figure 3-16, the end of RAM (25F$_H$) is saved in I/O locations SPH (3E$_H$) and SPL (3D$_H$). The variable temp is loaded with the value FF$_H$.

Next, the subroutine manipulate_temp is called. Figure 3-17 shows the changes in memory contents.

The contents of the variable temp (FF$_H$) are pushed on stack (e.g., RAM locations 25C$_H$ and 25D$_H$), the stack pointer is adjusted to 25C$_H$, and register temp is loaded with the new value A5$_H$.

In Figure 3-18, a manipulation of the new contents of register temp is simulated. Figure 3-19 shows the restored conditions after popping the temporarily stored original contents of variable temp.

As Figure 3-19 shows, the original contents of register temp and of the stack pointer are restored for further processing.

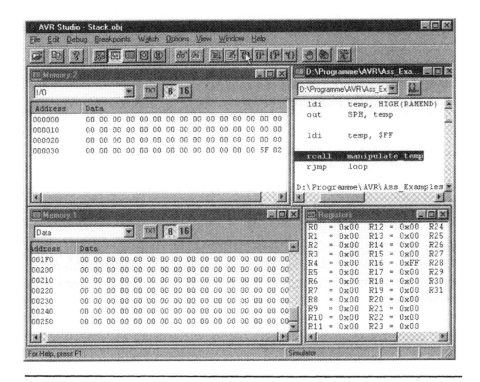

Figure 3-16
Stack pointer initalization simulated in AVR studio.

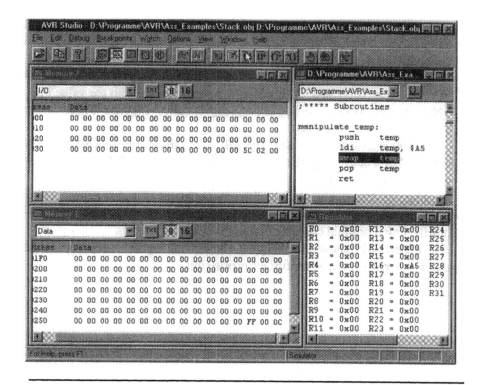

Figure 3-17
Pushing a word to stack simulated in AVR studio.

3.6 Program Constructs

High-level languages have constructs that support effective programming.
Programming in Assembler supports these constructs indirectly. This chapter
gives some examples of program constructs and their realization on the Assembler level.

3.6.1 Conditional Branches

A conditional branch means a branch in the program whose flow depends on
whether a condition is fulfilled or not. In high-level languages there are constructs such as IF ... THEN ... ELSE that are not available in Assembler.

AVR RISC Microcontroller Handbook

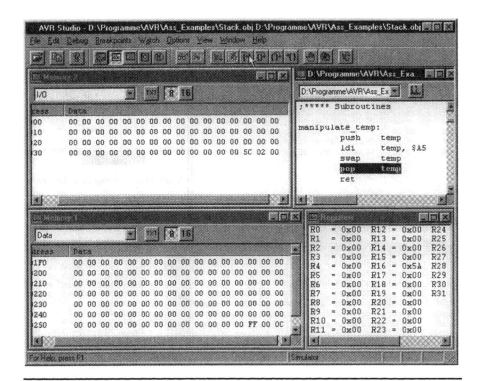

Figure 3-18
Manipulation of register contents simulated in AVR studio.

Figure 3-20 shows a structogram of a conditional branch. After the test of the condition we find two branches of the further program. The test itself is the test of one of the flags in the SREG register. It must be guaranteed by a previous operation that the flag will be influenced tested next.

If the condition was fulfilled, the program will operate the TRUE path. If the condition was not fulfilled, the program has to operate the FALSE path.

There are several instructions in each path. In the TRUE path we find the instructions T1 to Tn and in the FALSE path F1 to Fm. The index n or m means that the branches can have a different number of instructions.

From the description of the AVR instruction set, we know many branch and skip instructions. These instructions are suitable for testing the condition.

A very simple example program will explain conditional branching on the Assembler level.

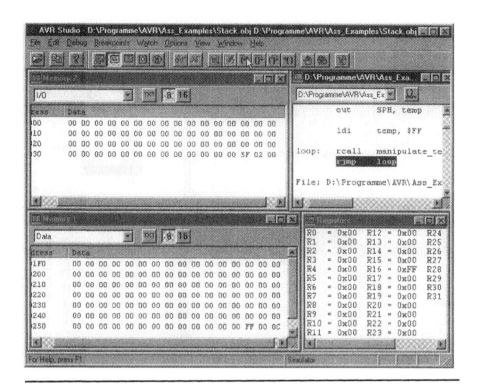

Figure 3-19
Popping a word from stack simulated in AVR studio.

```
          ldi     r16,$10      ; Load registers immediately
          ldi     r17,$20

          cp      r17,r16      ; Compare registers contents
          breq    equal        ; Branch if equal
not_equal:                     ; False path
          nop
          rjmp    eob          ; Skip the true path

equal: nop                     ; True path

eob:   nop                     ; End of branches
```

At the beginning, the registers R16 and R17 are loaded with any values (R16 = $10, R17 = $20). The third instruction compares both registers. As a result of this compare, some flags are set. The Z flag will be tested by the next instruction.

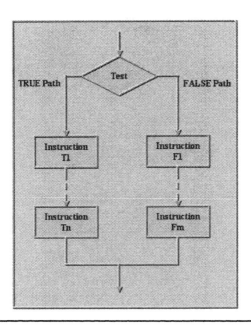

Figure 3-20
Conditional branch.

If the Z flag is set, then a branch to label `equal` occurs. The label `equal` opens the TRUE path of our conditional branch. The TRUE path in our example contains only one `nop` instruction before the end of the branch is achieved.

Otherwise, if the Z flag is cleared, a branch to label `not_equal` occurs. The label `not_equal` opens the FALSE path of our conditional branch. The FALSE path in our example contains only one `nop` instruction, for example. At the end of the FALSE path the `rjmp eob` instruction skips the TRUE path.

It is important to pay attention to this peculiarity. In Figure 3-20 TRUE and FALSE paths were in parallel, but programming is a linear act. Therefore, the notification of the conditional branch will differ slightly from its more general representation in the structogram.

3.6.2 Program Loops

Parts of the program that will operate several times, one after the other, are called program loops. Transmitting of a defined number of characters via a serial interface or polling a port pin several times are examples of such loops.

The examples already show that there are basic differences. The first kind of loop has a defined number of passes through the loop. Let us name this

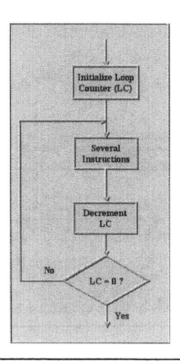

Figure 3-21
Defined loop.

kind of loop a defined loop. In the second kind of loop the number of passes is undefined and results from an operation in the loop. In contrast to defined loops, we name this second type of loop undefined loop.

Defined Loops In a defined loop, the number of passes is known before entering the loop body. Therefore, a variable working as a loop counter must be initialized at this point.

In the loop there are several instructions that will work as often as the loop counter allows. Mostly the loop counter acts as a down-counter so that after some decrements a test to zero can abort the loop. Figure 3-21 shows this construct of a defined loop in a structogram.

The following program lines shows the implementation of a defined loop in Assembler. Before the entry in the loop at label loop, the counter is initialized with the number of loop cycles (here $0A_H$). All instructions in the loop body are executed that number of times.

```
ldi    counter, $0a ; Counter defines the number
                    ; of going through the loop

loop:  nop                ; Any instruction(s) in the loop body
       dec   counter
       brne  loop

       nop                ;Further instructions out of the loop
```

In our example we found three instructions in the loop body—an optional nop instruction, the decrementation of the counter variable, and the conditional branch. This branch instruction tests the zero flag. The branch occurs depending on the test result.

If the zero flag is cleared, a branch to label loop happens. If the counter variable is decremented to zero, then the zero flag is set and no branch occurs. No branch means the execution of the next instruction out of the loop.

Undefined Loops In contrast to a defined loop, the exit in an undefined loop is dependent on the result of an internal operation. Figure 3-22 shows the structogram of an undefined loop.

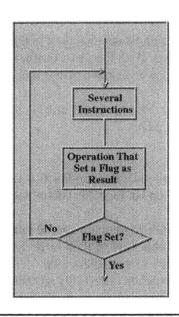

Figure 3-22
Undefined loop.

All instructions in the loop work one after the other repeatedly as long as the exit condition is not fulfilled. The exit condition is a flag set as result of an operation.

The following program lines show the implementation of an undefined loop in Assembler as often used for querying an input line(s) or status flag.

In the first program example, a 0-terminated ASCII string is stored in EEPROM beginning at location text1. At label start, this EEPROM location is addressed.

The EEPROM contents are read by the loop until the terminating character $00 is read and stored in register EEdrd_s temporarily.

```
start:ldi    temp, text1-1
      out    EEAR,temp       ; Set address of string

loop: rcall  EERead_seq      ; get EEPROM data
      out    PORTB, EEdrd_s  ; write to PortB

      rcall  DELAY           ; Wait a moment
      tst    EEdrd_s         ; last character?
                             ; (0-terminated string)
      breq   start
rjmp  loop                   ; Repeat endless
```

The second program example demonstrates polling an input line and waiting for a defined level on it. With the sbic instruction, a single bit in an I/O register can be tested for Lo level and a skip if it occurs.

```
read_pin:sbic PINB, PB0 ; Wait until Pin0 on PortB is Lo
         rjmp read_pin
         ret
```

This subroutine will be left once Pin0 on PortB is Lo.

Notes: If the pin stays on Hi, then this subroutine runs endlessly.

Endless Loops An endless loop is a special kind of loop. Even in microcontroller programming, this kind of loop turns up again and again. Figure 3-23 shows the structogram of an endless loop.

Each microcontroller program is basically an endless loop. After a program start by power-on reset, for example, normally the initialization of variables and/or registers is carried out, and after that the program works endlessly—repeating each program function one after the other.

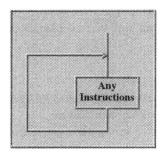

Figure 3-23
Endless loop.

The implementation of such an endless loop is very easy as the following listing shows. The nop instruction serves as a place-holder only. In a real application program it would replaced by the whole functionality of the program repeated endlessly.

```
endless:    nop         ; Placeholder for other instructions
            rjmp end
```

3.7 Defensive Programming

Defensive programming is one way to increase a microcontroller's immunity performance greatly. The beauty of defensive programming is that it is inexpensive to implement and, if done correctly, can save hardware costs in PCB layout as well.

3.7.1 Refreshing Port Pins and Important Registers

One of the simplest examples of defensive programming is a continous update of I/O pins and important registers.

In most microcontroller designs, the I/O pins generally will be laid out near the bond pads, which are consequently near the edge of the die. When the microcontroller is subjected to noise of a certain amplitude, the noise will propagate through to the silicon from the die edges inward. This means that logic at the edge of the die is most vulnerable to external noise sources such as the

I/O circuitry. So, by simply updating the data register and data direction registers at regular intervals, the threat of malfunctions can be reduced.

3.7.2 Polling Inputs

Another idea is to filter noise on input pins by reading the pin several times and taking the read value into consideration.

A typical application of a keyboard is polled in software and then read again several times within a 10-ms period to ensure that a true keypress has taken place. This type of polling is known as debounce protection.

If there is no mechanical bouncing, high-frequency polling can enhance security of an input query. The next excerpt of a program example shows such a polling subroutine with its internal logic.

```
.include "1200def.inc"

                rjmp    main
read_key:
not_a_one:      sbic    PINB, PB0
                rjmp    not_a_zero
                sbic    PINB, PB0
                rjmp    not_a_zero
                sbic    PINB, PB0
                rjmp    not_a_zero
                clc
; Carry is cleared after reading a 0 three times
                ret


not_a_zero:     sbis    PINB, PB0
                rjmp    not_a_one
                sbis    PINB, PB0
                rjmp    not_a_one
                sbis    PINB, PB0
                rjmp    not_a_one
                sec
; Carry is set after reading a 1 three times
                ret

main:           rcall   read_key
                ; value of pin is stored in carry flag
                ; further instructions follow
```

AVR RISC Microcontroller Handbook

The subroutine in the preceding example reads Pin0 of PortB in quick succession. If three reads of this pin are the same, the subroutine will be left with the carry flag at the state read.

A drawback of this polling approach is a possibly longer delay of this kind of query in a noisy environment. In time-critical applications, this could be a killer criterion.

A statistical approach for an input query can give the result in a defined time and could also be a solution for noisy environments.

The following excerpt of a program example contains a changed subroutine read_pin.

```
.include "1200def.inc"

.def samples   = r16
.def count     = r17

                rjmp    main

read_pin:       clr     samples
                ldi     count, 7
read_again:     sbic    PINB, PB0
                inc     samples
                dec     count
                brne    read_again
                cpi     samples, 4
                ret

main:           rcall   read_pin
                ; Test Carry for Input Level
                ; C=1 for 0,1,2,3
                ; C=0 for 4,5,6,7
                ; further instructions
```

The input pin is queried seven times. Each time a 1 is read, register samples are incremented. Before return from subroutine this count is compared to the middle of the range (here $(7+1)/2$). The Carry flag knows the result of that compare for making a decision afterwards.

Development Tools 4

The success of a new microcontroller family depends not only on the hardware concept itself, but also on the development tools supporting that microcontroller.

The development tools now available for the AVR microcontroller family include the following:

• ATMEL AVR Assembler
• ATMEL AVR Simulator
• IAR ANSI C-Compiler, Assembler, Linker, Librarian, & Debugger
• AVR Pascal from E-LAB Computers
• AVR BASIC from Silicon Studio
• ATMEL In-Circuit Emulator (ICE)
• EQUINOX Micro-Pro AVR Device Programmer

There are many development tools under development; please contact ATMEL for more details.

4.1 ATMEL AVR Assembler and Simulator

ATMEL supports the first steps with this new microcontroller with free development software consisting of ATMEL's AVR Assembler and Simulator. You will find this software on ATMEL's CD or on ATMEL's Web site for free download.

Figure 4-1 shows a screenshot of this software download possibility via Netscape. ATMEL's URL is listed in the Appendix. This Web site is also very helpful because it contains a growing number of application notes.

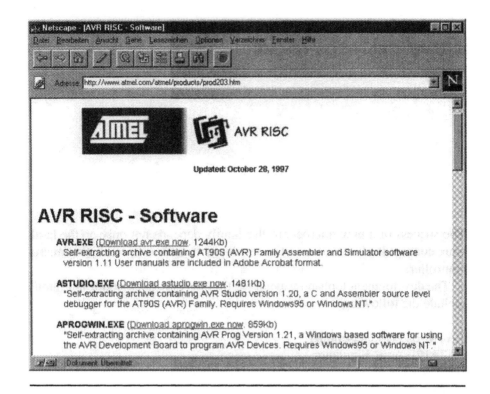

Figure 4-1
Download of AVR development tools from Atmel's Web site.

Figure 4-1 shows three downloadable archives:

- AVR.EXE—containing AVR Assembler and Simulator, including user manuals in PDF format
- ASTUDIO.EXE—containing the C and Assembler source-level simulator AVR Studio for Windows95/NT
- APROGWIN.EXE—containing the AVR Programmer Software for using the AVR Development Board to program the AVR microcontroller

Some further tools supporting the DOS environment can be found on this Web site.

After downloading and and running the self-extracting archives, you can start with the installation of the AVR Assembler and Simulator. That is all to be prepared for the development of the first Assembler program for the new AVR microcontroller family.

Installation of the development tools from ATMEL's CD is the easiest way to be prepared for the next steps. ATMEL's CD contains the ATMEL AVR Assembler and Simulator, as well as complete documentation of the members of the AVR microcontroller family.

The installation of the ATMEL AVR Assembler and Simulator is, therefore, a part of the installation procedure of the whole CD. Figure 4-2 shows that subdirectory containing the required setup program for an installation under Microsoft Windows. With a double-click on the setup icon, the well-documented installation procedure starts.

The ATMEL AVR Assembler and Simulator run under Microsoft Windows 3.11, Microsoft Windows95, and Microsoft Windows NT. In addition, there is an MS-DOS command-line version of the AVR Assembler. The Windows version of the AVR Assembler includes a full editor for writing Assembler programs.

In their Windows versions, both programs contain a valuable on-line help function covering most of the documentation.

4.1.1 ATMEL AVR Assembler

Like every assembler, the ATMEL AVR Assembler translates assembly source code into object code. The generated object code can be used as input to a simulator such as the ATMEL AVR Simulator or an emulator such as the ATMEL AVR In-Circuit Emulator.

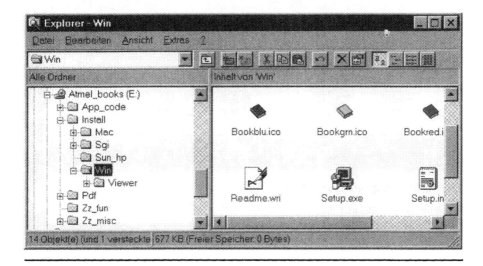

Figure 4-2
Installation of AVR Assembler and Simulator from ATMEL's CD.

The Assembler also generates a PROMable code that can be programmed directly into the program memory of an AVR microcontroller. Furthermore, the ATMEL AVR Assembler generates fixed code allocations; consequently, no linking is necessary.

Editing Assembler Source Code   At first we have a look to the source code in the file tutor1.asm. For this purpose we start the ATMEL AVR Assembler, and the Assembler window will open. Figure 4-3 shows the opened windows.

To load the file tutor1.asm, select this file via the menu **File>Open**. The loaded file will be listed in the Edit window (in the upper half of Figure 4-3). The assembling process can be started after loading and possibly inspection via the menu **Assemble**.

Immediately after the start of the assembling operation, the Message window (in the lower half of Figure 4-3) will open and document the operation. Note that this file contains errors for demonstration.

Some features of the ATMEL AVR Assembler are selectable by options. Over the menu **Options...**, the Options window can be opened. Figure 4-4

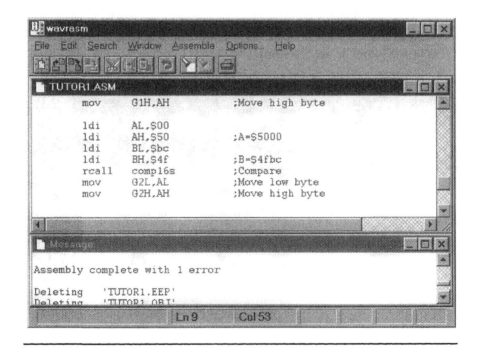

Figure 4-3
ATMEL AVR Assembler.

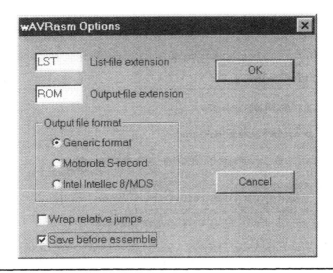

Figure 4-4
Options window.

shows this opened window and the options selectable by the user. The options are self-explanatory.

If you are working within the Assembler and have any questions, remember the powerful on-line help function.

Figure 4-5 shows the opened Help window. You will find information and examples of all aspects of programming with the ATMEL AVR Assembler.

Directives The source code itself consists of Assembler instructions and directives. The whole instruction set has been explained in detail already; therefore, we need only discuss the directives.

Table 4-1 shows a summary of all directives availabe for the ATMEL AVR Assembler. All directives listed in the table control the operation of the Assembler. To make a visible difference between directives and instructions in the Assembler source code, all directives must be preceded by a period.

There are directives for different purposes. A first group (BYTE, DB, DW, CSEG, DSEG, ESEG, DEF, EQU, SET, ORG) handles some memory and naming aspects.

A second group (LIST, LISTMAC, NOLIST) controls the list file generation, and a third group consists of more specialized directives such as MACRO, ENDMACRO, DEVICE, INCLUDE, and EXIT.

A documented example program, shown in Listing 4-1, demonstrates the use of directives.

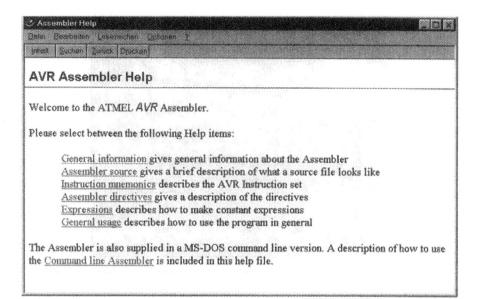

Figure 4-5
ATMEL AVR Assembler on-line help.

Table 4-1
Summary of directives.

Directive	Description
BYTE	Reserve byte to a variable
CSEG	Code segment
DB	Define constant byte(s)
DEF	Define a symbolic name on a register
DEVICE	Define which device to assemble for
DSEG	Data segment
DW	Define constant word(s)
ENDMACRO	End macro
EQU	Set a symbol equal to an expression
ESEG	EEPROM segment
EXIT	Exit from file
INCLUDE	Read source from another file
LIST	Turn listfile generation on
LISTMAC	Turn macro expansion on
MACRO	Begin macro
NOLIST	Turn listfile generation off
ORG	Set program origin
SET	Set a symbol to an expression

AVR RISC Microcontroller Handbook

```
;***********************************************************
;* Directive Test
;***********************************************************
;

.DEVICE AT90S8515          ; Prohibits use of non-implemented
                           ; instructions
.NOLIST                    ; Disable listfile generation
.INCLUDE "8515def.inc"     ; The included files will not be shown
                           ; in the listfile
.LIST                      ; Reenable listfile generation
        rjmp    RESET      ; Reset Handle
;***********************************************************
.EQU    tab_size=10        ; Set tab_size to 10
.DEF    temp=R16           ; Names R16 as temp
.SET    io_offset=0x23     ; Set io_offset to 0x23
;.SET   porta=io_offset+2  ; Set porta to io_offset+2
                           ; (commented because defined
                           ; in 8515def.inc)

.DSEG                      ; Start data segment
table:    .BYTE tab_size   ; reserve tab_size bytes in SRAM

.ESEG                      ; Start EEPROM segment
eeconst:.DB 0xAA, 0x55     ; Defines constants

.CSEG                      ; Start code segment
RESET:    ser    temp      ; Initializes temp (R16) with $FF
          out    porta,temp  ; Write contents of temp to Port A
          ldi    temp,0x00 ; Load address to EEPROM address
                           ; register
          out    EEAR,temp
          ldi    temp,0x01 ; Set EEPROM read enable
          out    EECR,temp
          in     temp,EEDR ; Read EEPROM data register

          clr    r27       ; Clear X hi byte
          ldi    r26,0x20  ; Set X lo byte to $20
          st     X,temp    ; Store temp in SRAM

forever:rjmp    forever    ; Loop forever
```

Listing 4-1
Use of directives.

Very important for the ATMEL AVR Assembler are its powerful macro capabilities. Therefore, a separate section is reserved for macros.

Macros Macros enable the user to build a virtual instruction set from normal Assembler instructions. You can understand a macro as a procedure on the Assembler instruction level.

The .MACRO directive tells the Assembler that this is the start of a macro and takes the macro name as its parameter. When the name of the macro is written later in the program, the macro definition is expanded at the place it was used. A macro can take up to 10 parameters. These parameters are referred to as @0–@9 within the macro definition. When a macro call is issued, the parameters are given as a comma-separated list. The macro definition is terminated by an .ENDMACRO directive.

A simple example will show the definition and use of a macro:

```
.MACRO SUBI16              ; Start macro definition
       subi @1,low(@2)     ; Subtract low byte
       sbci @0,high(@2)    ; Subtract high byte
.ENDMACRO                  ; End macro definition

.CSEG                      ; Start code segment
SUBI16 r17,r16,0x1234      ; Sub.0x1234 from r17:r16
```

A macro to subtract immediate a word from a double register named SUBI16 is defined. In two steps, Hi-byte and Lo-byte are subtracted from this double register. Subtracting the Hi-byte takes the carry flag into consideration.

When the macro SUBI16 is called, the symbolic parameters (@0, @1, @2) are replaced by registers or immediate data.

By default, only the call to the macro is shown on the listfile generated by the Assembler. In order to include the macro expansion in the listfile, a LISTMAC directive must be used. A macro is marked with a + in the opcode field of the listfile.

The following excerpt from the listfile of the example program shows this detail:

```
. . .
            .MACRO SUBI16              ; Start macro definition
                subi @1,low(@2)        ; Subtract low byte
                sbci @0,high(@2)       ; Subtract high byte
            .ENDMACRO                  ; End macro definition

000001 e112    RESET:ldi r17,0x12
000002 e304          ldi r16,0x34
000003    +    SUBI16 r17,r16,0x1234 ; Sub.0x1234 from
               r17:r16

. . .
```

The macro definition consists of two subtraction instructions. Before a call of this defined macro is possible, the double register must be loaded. These operations will be done by the two `ldi` instructions. After that, the macro call calculates the chosen subtraction.

4.1.2 ATMEL AVR Simulator

The ATMEL AVR Simulator covers the whole range of microcontrollers in the AVR family. It executes object code generated for the AVR microcontrollers. In addition to being an instruction set simulator, it supports simulation of various I/O functions. ATMEL's AVR Assembler generates a special defined object file format readable by the ATMEL AVR Simulator. This object format allows Assembler source-level simulation.

The ATMEL AVR Simulator runs under Microsoft Windows 3.11, Windows95, and Windows NT. Figure 4-6 shows the user interface of ATMEL's AVR Simulator.

In the opened Simulator window, five windows are placed. In the top left window, the Assembler source code is listed. At top right, the Registers window is opened. All register contents are displayed for evaluation. To inspect

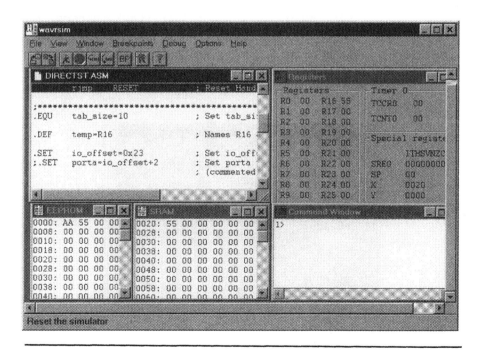

Figure 4-6
ATMEL AVR Simulator.

the contents of EEPROM and SRAM, see the EEPROM and SRAM windows at bottom left and middle. To control the simulator, at bottom right the Command window is opened.

For correct simulations, the microcontroller-dependent options must be set via the menu **Options>AVR Options**. Figures 4-7 and 4-8 show the options for two different types of AVR microcontrollers. The AT90S1200 operates with a three-level hardware stack, whereas the AT90S8515 uses SRAM as stack. If you want to simulate programs using stack operations, this difference must be noted.

During interrupts and subroutine calls, the return address program counter is stored on the stack. Except for the AT90S1200, the stack of the other AVR microcontrollers is allocated to the general data SRAM. Consequently, the stack size is only limited by the total SRAM size and the usage of the SRAM.

All user programs must initialize the stack pointer in the reset routine (before subroutines or interrupts are executed). The 16-bit stack pointer SP is read/write accessible in the I/O space.

The required initialization for microcontrollers other than the AT90S1200 is very easy, as the following source code lines show. The stack pointer is loaded here with the last memory cell in SRAM.

```
RESET:  ldi    temp, low(RAMEND)
        out    SPL, temp            ; Init Stack Pointer Lo
        ldi    temp, high(RAMEND)   ; For SRAM > 256 byte
        out    SPH, temp            ; Init Stack Pointer Hi
```

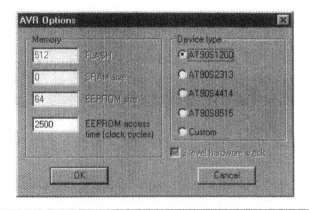

Figure 4-7
AVR options for AT90S1200.

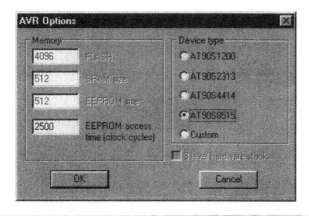

Figure 4-8
AVR Options for AT90S8515.

The stack pointer is decremented by 1 when data is pushed onto the stack with the PUSH instruction, and it is decremented by 2 when data is pushed onto the stack with a subroutine call or an interrupt. The stack pointer is incremented by 1 when data is popped from the stack with the POP instruction, and it is incremented by 2 when data is popped from the stack with return from subroutine RET or return from interrupt RETI. So, the stack always grows to lower addresses in SRAM.

The ATMEL AVR Simulator can be controlled through the Command window, through menus, or by clicking on toolbar buttons. Using the toolbar buttons very comfortable debugging is possible. Figure 4-9 shows the **Debug** menu. For each debugging function, a toolbar button exists.

Working with the simulator is well supported by an on-line help system. The on-line help is organized as a hypertext system, letting you find the required help very easily after opening the help system via menu **Help>Option**. Figure 4-10 shows an explanation of the toolbar buttons.

In most debugging situations, working with the toolbar buttons gives excellent support for the debugging task. Enhanced features are reached by use of simulator commands. These commands, which are not often used, are also explained by the on-line help system. Figure 4-11 shows the explanation of simulator commands in the on-line help system.

The use of simulator commands was explained by a short source-code excerpt presented earlier. The following source code reads EEPROM address 0x00 and stores the contents in SRAM address 0x20.

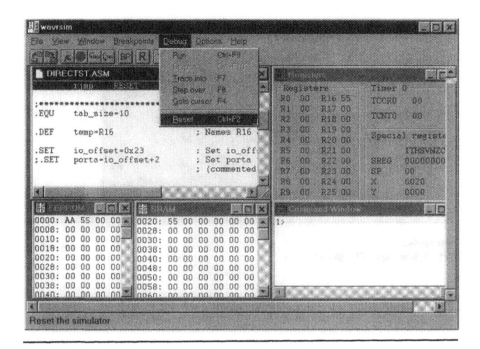

Figure 4-9
Debug options.

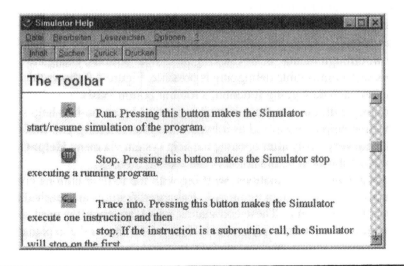

Figure 4-10
Explanation of toolbar buttons.

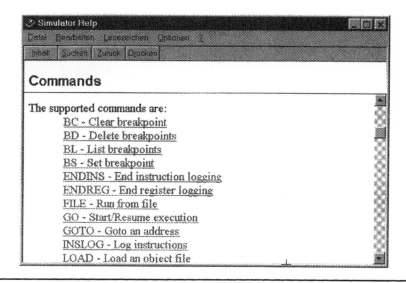

Figure 4-11
Explanation of commands.

```
ldi       temp,0x00      ; Load address to EEPROM address
                         ; register
out       EEAR,temp
ldi       temp,0x01      ; Set EEPROM read enable
out       EECR,temp
in        temp,EEDR      ; Read EEPROM data register

clr       r27            ; Clear X hi byte
ldi       r26,0x20       ; Set X lo byte to $20
st        X,temp         ; Store temp in SRAM
```

The contents of both first EEPROM cells are 0xAA and 0x55, as Figure 4-12 shows. Our short program would read from cell 0x00 the contents 0xAA and store this value in SRAM.

In our example, we want to store not the contents of cell 0x00, but the contents of cell 0x01, without reassembling.

After pressing the Reset button, we can step to the source code line marked in Figure 4-12 by pressing the Step button in the toolbar. One instruction before the marked instruction, the register temp (= R16) was loaded with 0x00 for addressing the EEPROM cell 0x00. To avoid an access to cell 0x00, we have to overwrite register R16 with 0x01. This operation is executed by typing the command reg 16,1, followed by Enter.

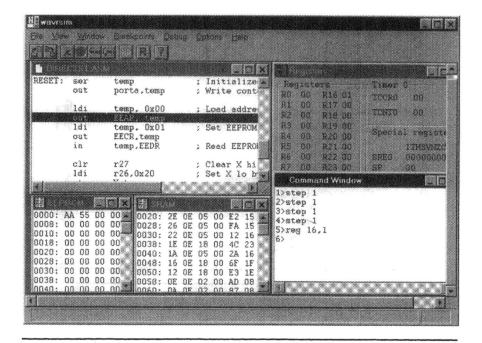

Figure 4-12
Manipulation of register contents.

The Command window shows the typed command. In the Register window the changed content of register R16 is visible.

We can now expect, after further stepping, an access to EEPROM cell 0x01 and the storage of value 0x55 to SRAM. Figure 4-13 shows this result in the SRAM window. In cell 0x20, the contents of EEPROM cell 0x01 are now stored.

This simple example of working with the ATMEL AVR Simulator should show some of its features.

4.2 ATMEL AVR Studio

AVR Studio enables the user to fully control execution of programs on the AVR In-Circuit Emulator or on the built-in AVR Instruction Set Simulator. AVR Studio supports source-level execution of Assembly programs assembled with Atmel's AVR Assembler and C programs compiled with IAR's ICCA90 C Compiler for the AVR microcontrollers. AVR Studio runs under Microsoft Windows95/NT.

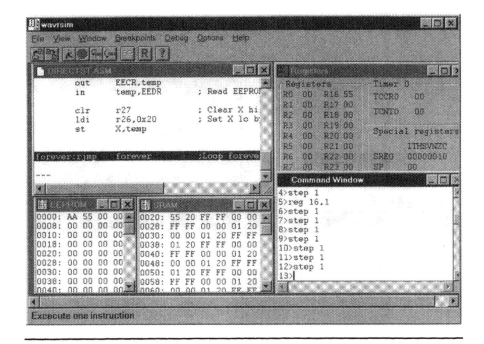

Figure 4-13
Changed EEPROM access.

AVR Studio can be targeted toward an AVR In-Circuit Emulator or the built-in AVR Simulator. When the user opens a file, AVR Studio automatically detects whether an Emulator is present and available on one of the system's serial ports.

If an Emulator is found, it is selected as the execution target. If no Emulator is found, execution will be done on the built-in AVR Simulator instead. The Status bar will indicate whether execution is targeted at the AVR In-Circuit Emulator or the built-in AVR Simulator.

Figure 4-14 shows the simulation of an Assembler program in AVR Studio.

The user has full control of the status of each part of the microcontroller using many separate windows:

- Register window displays the contents of the register file
- Watch window displays the values of defined symbols (in C programs)
- Message window displays messages to the user
- Processor window displays information such as Program Counter, Stack Pointer, Status Register, and Cycle Counter
- Memory windows show program, data, I/O, and EEPROM
- Peripheral windows show 8-bit timer, I/O ports, and EEPROM registers

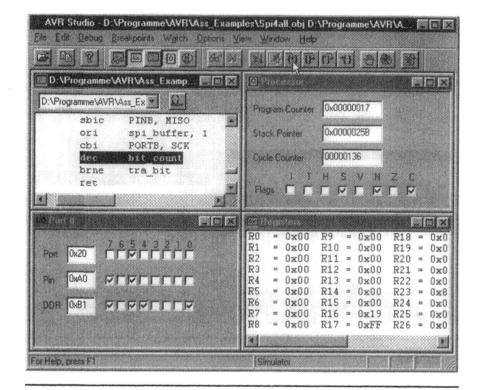

Figure 4-14
Simulation of an Assembler program in AVR Studio.

AVR Studio supports all available types of AVR microcontroller. The setup of the right device happens over the menu **Options > Simulator Options**. Figure 4-15 shows the window for setting up the Simulator options.

The simulation with AVR Studio itself is more comfortable than simulation with the AVR Simulator, although the handling is quite similar. A detailed description of the simulation seems less important at this point.

4.3 IAR Embedded Workbench EWA90

The IAR Embedded Workbench is a highly evolved development tool for programming embedded applications. The tool offers the choice of C to all AVR microcontroller applications, from single-chip to banked design. With its built-in chip-specific optimizer, the compiler generates very efficient, fast, and reliable PROMable code for the AVR derivatives.

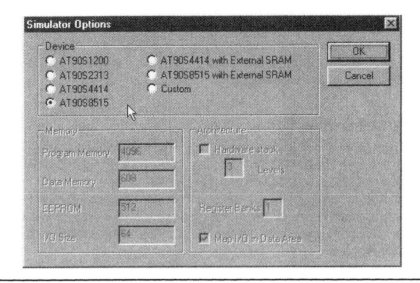

Figure 4-15
AVR Studio—Simulator options.

The EWA90 takes full advantage of the 32-bit Windows95/NT environment. The toolset can also run under Windows 3.11 with the Win32s subsystem, included in the software package. The EWA90 implements the intuitive Windows95 interface with all its features. Furthermore, the EWA90 integrates the IAR C compiler, linker, librarian, and assembler in a seamless environment with an easy-to-use project feature and option handling.

The Make system automatically generates a dependency list of output files, source files, and even include files. This allows the Make system to recompile or reassemble only the updated parts of the source code, which speeds up the building process.

The EWA90 offers flexibility in terms of customizable toolbar and user-defined shortcut keys. The Editor implements the basic Windows editing commands as well as extensions for C programming, such as C syntax coloring, and direct jump to context from error listing.

The on-line help function makes it easy to quickly find specific help about the tool without leaving the Embedded Workbench, which reduces your learning time and, thus, time to market.

The IAR C compiler is also available under DOS and is then called ICCA90. The DOS version comes with a mouse-controlled, menu-driven user interface, allowing all development steps to be performed in an integrated DOS environment.

4.3.1 Summary of Available AVR Tools

The available tools for AVR microcontrollers are listed in Table 4-2. For the AVR microcontrollers explained here, the baseline versions of EWA90 and CWA90 are completely sufficient.

4.3.2 IAR C Compiler

The IAR C compiler, the core product in the EWA90 and the ICCA90, is fully compatible with the ANSI C standard. As Table 4-3 shows, all data types required by ANSI are supported without exception.

Full ANSI C compatibility also means that the compiler conforms to all requirements placed by ANSI on run-time behavior, even those less-known yet important requirements such as integral promotion and precision in calculating floating-point.

Float and Double are represented in IEEE 32-bit precision. Struct, array, union, enum, and bitfield are also supported.

The compiler generates fully reentrant code. Any function can be interrupted and called from the interrupting routine without the risk of corrupting the local environment of the function. This feature makes the IAR C compiler ideal to use with real-time operating systems. Recursion is also supported.

Depending on the application, execution speed may be more critical than smaller code. To meet this need, the compiler has the powerful feature of allowing the user to favor speed optimization over code size.

Table 4-2
Available tools for AVR microcontrollers.

EWA90	Windows Embedded Workbench, including Compiler, Assembler, and Linker, as well as command-line support
EWA90 Baseline	A limited version of EWA90 with a maximum code size 8 kb and no command-line or floating-point support; suitable for the baseline derivatives of the AVR microcontroller
CWA90	Debugger/Simulator for the EWA90
CWA90 Baseline	A limited version of CWA90, suitable for the baseline derivatives of the AVR microcontroller (to be used with the EWA90 baseline)
ICCA90	Integrated C compiler for DOS, including Assembler and Linker
CSA90	Debugger/Simulator for the ICCA90
AA90	Assembler

AVR RISC Microcontroller Handbook

Table 4-3
Data types.

Data Type	Size in Bytes	Value Range
sfrb	1	0 to 255
sfrw	2	0 to 65535
signed char	1	-128 to +127
unsigned char	1	0 to 255
signed short	2	-32768 to +32767
unsigned short	2	0 to 65535
signed int	2	-32768 to +32767
unsigned int	2	0 to 65535
signed long	4	-2^{31} to 2^{31}-1
unsigned long	4	0 to 2^{32}-1
float IEEE 32 bit	4	±1.18E-38 to ±3.39E+38, 7 digits
pointer	1-2	object address

To ideally suit development for embedded systems, standard C needs additional functionality. IAR Systems has defined a set of extensions to ANSI C, specific to the AVR architecture. Table 4-4 lists all of these extended keywords. They can be invoked by using the #pragma directive, which maintains compatibility with ANSI and code portability.

It is possible to access specific memory locations directly from C. The following example shows how locations 0x10 and 0x11 are accessed:

```
sfrb P0IN  = 0x10;
sfrb P0OUT = 0x11;

void read_write(char c)
{
        POUT = c;    /*writes c to location 0x11*/
        c = POIN;    /*reads location 0x10 into c*/
}
```

All these extensions, coupled with absolute read/write access, minimize the need for assembly-language routines.

The compiler comes with full floating-point support. It follows the IEEE 32-bit representation using an IAR Systems proprietary register-based algorithm, which makes floating-point manipulation extremely fast.

Table 4-4

IAR Systems embedded C extensions.

Type	Keyword	Description
Function	interrupt	Creates an interrupt function that is called through an interrupt vector. The function preserves the register contents and the processor status.
	monitor	Turns off the interrupts while executing a monitor function.
	C-task	Declares a function as not callable (e.g., main) to save stack.
Variable	no_init	Puts a variable in the no_init segment (battery-backed RAM).
	sfrb	Maps a byte value to an absolute address.
	sfrw	Maps a word to an absolute address.
	tiny	Access using 8-bit address.
	near	Access using 16-bit address.
	flash	Access in the program address space.
Segment	codeseg	Renames the CODE segment.
	constseg	Creates a new segment for constant data.
	dataseg	Creates a new DATA segment.
		(These are mostly used to place code and data sections in nonconsecutive address ranges.)
Intrinsic	_SEI	Enable interrupt.
	_CLI	Disable interrupt.
	_NOP	NOP instruction.
	_OPC	Inserts the opcode of an instruction into the object code.
	_LPM	Returns 1 byte from the program address space.
	_SLEEP	Enter sleep mode.
	_WDR	Watchdog reset.

4.3.3 Macro-Assembler for Time-Critical Routines

The IAR C compiler kit comes with a new relocatable macro assembler. This provides the option of coding time-critical sections of the application in Assembler without losing the advantages of the C language. The preprocessor of the C compiler is incorporated in the Assembler, thus, allowing header files to be shared. C include files can also be used in an Assembler program. All modules written in Assembler can easily be accessed from C and vice versa, making the interface between C and Assembler a straightforward process.

The Assembler provides an extensive set of directives to allow total control of code and data segmentation. Directives also allow creation of multiple modules within a file, macro definitions, and variable declarations.

4.3.4 Linker

The IAR XLINK linker supports complete linking, relocation, and format generation to produce AVR PROMable code. XLINK generates more than 30 different formats and is compatible with most popular emulators and EPROM burners. The XLINK is extremely versatile in allocating any code or data to a start address and in checking for overflow. Detailed cross-reference and map listing with segments, symbol information, variable locations, and function addresses are easily generated.

4.3.5 ANSI C Libraries

The IAR C compiler kit comes with all libraries required by ANSI. Additionally, it comes with low-level routines required for embedded systems development. Table 4-5 lists these routines provided in source code.

4.3.6 IAR CWA90 Debugger/Simulator

The IAR C-SPY, CWA90, is a high-level language debugger incorporating a complete C expression analyzer and full C-type knowledge. It combines the detailed control of code execution needed for embedded development debugging with the flexibility and power of the C language. CWA90 shows the calling stack as well as tracing on both statement and Assembler levels. The source window can display C source code and mix it with Assembler. There is also a "locals" window showing the auto variables and parameters for the current function.

The optional CWA90 is integrated with the Embedded Workbench, implementing the intuitive Windows95 interface with all its features. CWA90 is user-friendly with customizable toolbar, drag and drop facility, and user-configurable shortcut keys. It is also configurable in the sense that the user can choose which windows to display.

CWA90 allows you to set an unlimited number of breakpoints. Breakpoints can be set on C statements, on Assembler instructions, and on any address with an access type of read, write, or opcode fetch, or as a combination of these. The breakpoint can be extended with an optional condition. After a breakpoint is triggered, any optional macro commands can be executed.

Table 4-5
Library functions.

Interrupt simulation implements commands to launch specific interrupts at a specific cycle count or periodically to a given cycle interval. The interrupt simulation can also be set to generate intermittent interrupts. The simulator then uses the same algorithm as the hardware for choosing the highest priority interrupt to be executed.

Breakpoint simulation and macro language allow most complex external environments to be simulated. Since I/O simulation is built-up with the macro language, it is easy to customize and very easy to extend. CWA90 terminal I/O emulation offers a console window for target system I/O. This unique feature is useful for debugging embedded applications when logical flows are of interest or the target is not yet ready.

The Watch window makes it possible to watch any expression. The window itself will be updated whenever a breakpoint is triggered or a step is finished. Any variable can be modified during execution by using specific C expressions.

The source window for the Assembler debugger displays the Assembler instructions. It has a built-in assembler and disassembler function, menu, and register window, and it can evaluate Assembler expressions.

4.3.7 EWA90 Demo of AVR Embedded Workbench

The EWA90 demo has almost full functionality, except that it will only produce scrambled UBROF debug format as output. No PROMable code is buildable with the EWA90 demo. There is also a limit to the size of the modules handled by the different tools.The following limitations apply to this demo version of the EWA90 product in terms of maximum code size:

• C compiler: 2 Kbytes
• Assembler: 1 Kbyte
• Linked application: 2 Kbytes
• Debugger: 2 Kbytes

The demo includes the following:

• The Embedded Workbench EWA90, which is an integrated project development environment in Windows and includes an ANSI C compiler, assembler, linker librarian, and run-time libraries, as well as an editor that offers flexibility in terms of a customizable toolbar and user-defined shortkeys (not only limited to the editor). The editor implements the basic Windows editing commands, as well as extensions for C programming, such as C syntax highlighting, and direct jumps to context from error listing. A project maintenance function makes it possible to have several targets and a debug target with different settings of options.
• The command-line (DOS) versions of the C compiler, Assembler, Linker, and Librarian.
• The C-SPY Debugger/Simulator, which combines the detailed control of code execution needed for embedded development debugging with the flexibility and power of the C language and includes features such as a C-like macro language, I/O and interrupt simulation, and setting of complex breakpoints.
• Extensive on-line help, which make it easy to find specific help about the tool without leaving the Embedded Workbench. It also includes an extensive tutorial to help you get started using the toolset.

IAR Systems offers this EWA90 demo for a free download from their Web site (see Figure 4-16). The URL of their Web site is listed in the Appendix.

To give a first impression of EWA90-based project handling, some screenshots will be used to demonstrate this comfortable environment.

After the IAR Embedded Workbench is started, the opening window gives a first impression of the resources available. Figure 4-17 shows this opening window.

Very important for an integrated environment are comfortable project handling and source text editing capabilities. Figure 4-18 shows the edit windows listing the first C program example, `HelloWorld.C`. This very simple program receives a character from RS-232. After it is received, the string "Hello World" will be sent to RS-232 followed by the string "The received character was." Finally, the received character will be echoed back to RS-232.

The result of compiling this first program example is shown in Figure 4-19. The successful compiler run is displayed in its own message window.

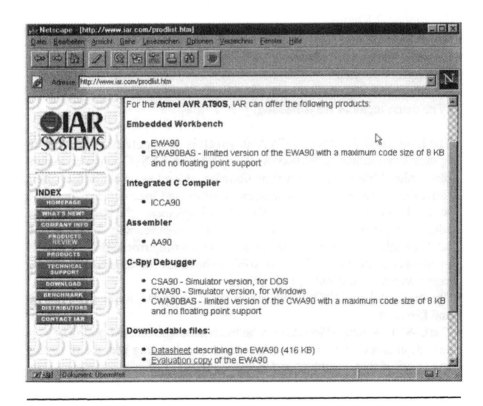

Figure 4-16
EWA90 demo download from the Web.

AVR RISC Microcontroller Handbook

Figure 4-17
Starting the EWA90.

If the compiler run was error-free, the simulation of the compiled program can be started by pressing the C-SPY button. The cursor in Figure 4-20 points to this button.

The simulator C-SPY presents in a typical layout. Figure 4-21 shows three opened windows during the simulation of HelloWorld.C.

Buttons for controlling the program flow (e.g., single-step operation, breakpoint setting) are available. C-SPY includes a window for terminal I/O. It is possible to show the resulting Assembler code beside the C source in the source window.

4.4 AVR Pascal from E-LAB Computers

The German company E-LAB Computers develops Pascal compilers supporting small microcontrollers as well as the AVR microcontroller.

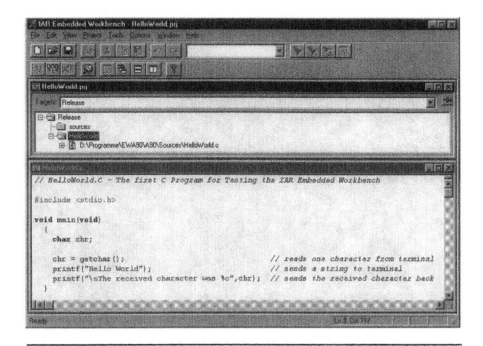

Figure 4-18
First test of IAR Embedded Workbench—`HelloWorld.C`.

E-LAB Computers' AVR Pascal compiler is embedded in the Windows-based program development environment PED32. Figure 4-22 shows the opening window of the programmers editor.

The bottom line displays some other microcontrollers supported by E-LAB Computers.

Figures 4-23 and 4-24 give an idea of project handling and source-text editing in the programming environment PED32. All required activities such as handling projects, opening and saving files, printing files, cut, copy and paste, compile, assemble, and simulate can be handled from the toolbar within PED32.

Some details of the Pascal language for programming the AVR microcontroller are described in the following tables. Table 4-6 shows the types available in AVR Pascal. In addition to the standard types, we can find enumeration, pointer, and some more-byte formats including floating-point.

The operators listed in Table 4-7 are more or less standard. Depending on the available memory, there can be some restrictions for strings and arrays. The IN operator together with enumeration is very useful, as the following program lines show.

```
Type  eKey = (Key1, Key2, Key3);       {Type Declaration}

Var   Keys : eKey;    {Variable Keys of type eKey is defined}
                      {with the elements Key1, Key2, and Key3}

If Keys in [Key1..Key3] then ..

If Keys = Key2 then . . .

If c in ['a'..'g'] then . . .
```

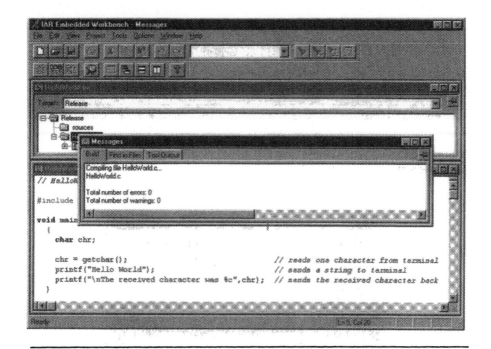

Figure 4-19
Result of compiling.

Figure 4-20
Starting C-SPY.

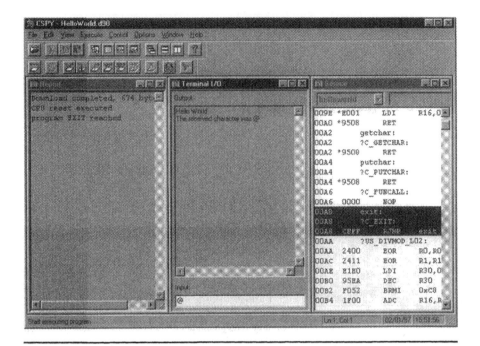

Figure 4-21
Simulation of `HelloWorld.C`.

Figure 4-22
Program editor opening window.

AVR RISC Microcontroller Handbook

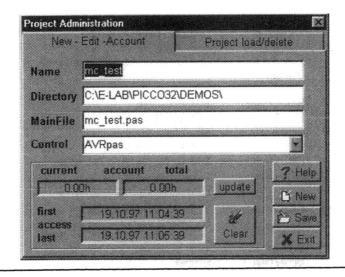

Figure 4-23
Project window in PED32.

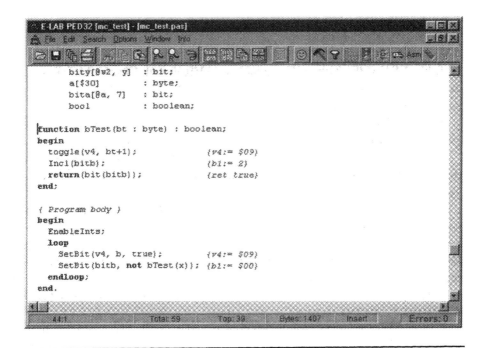

Figure 4-24
Edit window in PED32.

Table 4-6

Types in AVR Pascal.

Types	Comment
BOOLEAN	8bit, true..false
BYTE	8bit, 0..255
CHAR	8bit, chr(0..255)
BIT	1bit, true..false, 0..1
ENUM	8bit, 0..255, enumeration type
STRING	0..255 characters
	(limited as VAR in small CPUs)
ARRAY	255 characters maximum
	(limited as VAR in small CPUs)
PROCEDURE	16bit, parameter, word, address
WORD	16bit, 0..65535
POINTER	8/16bit
INTEGER	16bit, -32767..32768
LONGWORD	32bit, 0..4294967295
LONGINT	32bit, $-2^{31} .. 2^{31}$-1
FLOAT	24bit, $10^{-18}..10^{18}$

Table 4-8 lists the keywords available in AVR Pascal. Clearly AVR Pascal does not have a reduced instruction set and comfortable program structures can be built.

For effective programming, compilers have access to libraries. Tables 4-9 and 4-10 show the implemented library functions for AVR Pascal.

The AVR Pascal Compiler optimizes all library accesses: only functions that are really used will be imported and consume memory (code and data). This avoids a waste of resources.

Nearly all compilers build programs based on stack machines. All parameters and temporary results are handled over the stack. As a result, many unnecessary PUSH and POP operations will take place. Also, variables will be loaded in working registers even if they are already there. The run-time optimization recognizes and eliminates these operations.

After the introduction of AVR Pascal, a short view of the required source-text frame closes this section.

For formal reasons, the source text has to have a defined frame beginning with Program Name; and ending with End. The editor PED32 generates such a frame based on a template with each new project automatically. Listing 4-2 shows a suitable frame for AVR Pascal programs,

Table 4-7

Operators in AVR Pascal.

Operators	Comment	Example
Numerical and Logical Operators		
NOT	Logical NOT	`a:= not a;`
DIV	Division	`a:= div b;`
MOD	Modulo operation	`a:= a mod 5;`
AND	Logical AND	`a:= a and $0f;`
OR	Logical OR	`a:= a or $30;`
XOR	Logical XOR	`a:= a xor b;`
SHL	Shift left	`a:= a shl 5;`
SHR	Shift right	`a:= a shr b;`
ROL	Rotate left	`a:= a rol 4;`
ROR	Rotate right	`a:= a ror x;`
IN		`if ch in ['a'..'z'] then`
+	Addition	`a:= a + 5;`
-	Subtraction	`a:= a - b;`
*	Multiplication	`a:= a * b;`
/	Float division	`a:= a / 5.5;`
Compare Operators		
>	Greater than	`if a > b then`
<	Less than	`if a < b then`
=	Equal	`if a = b then`
>=	Greater or equal	`if a >= b then`
<=	Less or equal	`if a <= b then`
<>	Not equal	`if a <> b then`
Unary Operators		
@	Returns address	`x:= @a;`
^	Pointer dereference	`x:= p^;`
%	Binary constant	`x:= %01100101;`
$	Hexadecimal constant	`x:= $00FF;`
#	Decimal display of char	`x:= #13;`

Table 4-8
Keywords in AVR Pascal.

Keyword	Comment
PROGRAM	Beginning main program
DEVICE	Processor and hardware specification
IMPORT	Import of system functions
FROM	Reserved, not implemented yet
DEFINE	Parameter for import functions
IMPLEMENTATION	Beginning of program
TYPE	Type definition
CONST	Beginning of declaration of constants
VAR	Beginning of declaration of variables
PROCEDURE	Procedure declaration
FUNCTION	Function declaration
FORWARD	Forward declaration of functions or procedures
BEGIN	Beginning of body of procedures or functions
RETURN	Abort and exit of procedures and/or functions
GOTO	Absolute jump within a block
LABEL	Target for absolute jump within a block
END	End of body of procedure or function
ASM	Beginning of Assembler text
ENDASM	End of Assembler text
IF	IF Statement needs at least THEN and ENDIF
THEN	IF a > b then a:= b; EndIf;
ELSE	IF a > b then a:= b else b:= a; EndIf;
ELSIF	IF a > b then a:= b elsif b = a then inc(a) else dec(a); EndIf;
ENDIF	End of an IF statement
CASE	Beginning of a CASE Statement—Case x of
ENDCASE	End of a CASE Statement
FOR	Beginning of a FOR loop—For x:= 0 to 10 do
ENDFOR	End of a FOR loop
WHILE	Beginning of a WHILE loop
ENDWHILE	End of a WHILE loop
REPEAT	Beginning of a REPEAT loop
UNTIL	End of a REPEAT loop
BREAK	Abort of a FOR, REPEAT, or WHILE loop
LOOP	Beginning of an endless loop
ENDLOOP	End of an endless loop
EXITLOOP	Abort of an endless loop

Table 4-9
Standard library functions of AVR Pascal.

Standard Library Functions	Comment
TRUE	Predefined constant
FALSE	Predefined constant
NIL	Predefined constant for pointer
LO	Lower byte of a 2-byte value—a:= lo(i)
HI	Higher byte of a 2-byte value—a:= hi(i)
ABS	Absolute value of an integer or longint value
INCL	Bit set—Incl(a, 5);
EXCL	Bit clear—Excl(Leds, a);
TOGGLE	Bit toggle—Toggle(Leds, 4);
BIT	Bit test— if Bit(Eing1) then.. ; Endif;
SETBIT	Bit switch— SetBit(Eing1, true);
INC	Variable increment— inc(a);
DEC	Variable decrement—dec(b);
SWAP	Swap low nibble and high nibble for Byte and Char
	Swap low byte and high byte for Word and Integer
ODD	Test for odd value—if ODD(x) then.. ; Endif;
LENGTH	Returns the actual string length
SIZEOF	Returns the memory needs of an object
FILLBLOCK	Fills a memory area
COPYBLOCK	Copies a memory area
	Converts a number into string
BYTETOSTR	Write(LCDout, ByteToStr(v1:0:2) + 'msec ');
INTTOSTR	Write(LCDout, ' ' + IntToStr(V4:7:4));
	str:= IntToStr(1234:6:2);
BYTETOHEX	Converts a number into hex string
	str:= ByteToHex(b);
INTTOHEX	str:= IntToHex(1234);
WRITE	Writes a string or a number with conversion by a
	procedure or variable—Write(proc, 'x');
READ	Reads a character or string
MDELAY	Software delay in msec (1..65000)
SOUND	Frequency/sound output
UDELAY	Software delay in µsec (10..2560)
ENABLEINTS	Enables the global interrupt
DISABLEINTS	Disables the global interrupt
WATCHDOG	Triggers the watchdog

Table 4-10
Hardware-dependent library functions of AVR Pascal (excerpt).

Hardware Dependent Library Functions	Comment
PROCCLOCK	Processor clock in hertz
STACKSIZE	Expected stack size in bytes
RUNTIMEERR	Declaration and import of Range/StackCheck procedure
SYSTICK	Timer-controlled interrupt for timer functions
INTERRUPT	Declaration of an interrupt handler
FLAGS	Byte variable, contains some flags
TIMFLAG	Bit variable, set on each system tick
INTFLAG	Bit variable, copy of global interrupt enable
TIMER	Variable, decremented on each system tick
SWITCHP	Defines a port address for debounced port
PORT_STABLE	Variable for debounced port
INP_STABLE	Function (SwitchP): if INP_STABLE(1) then..
INP_RAISE	Function (SwitchP): if INP_RAISE(0) then..
PWMPORT	Defines pre-scaler and frequency for PWM ports 1 & 2
PWMPORT1	Definition and import of PWMport1
PWMPORT2	Definition and import of PWMport2
LCDPORT	Definition and import of a LCD port
LCDOUT	Write to LCD port
LCDINP	Read from LCD port
LCDCTRL	Write to LCD control port
LCDSTAT	Read LCD status port
SERPORT	Definition and import of serial interface
RXBUFF	Definition of RX buffer length
TXBUFF	Definition of TX buffer length
SERINP	Reads from serial interface or RX buffer
SEROUT	Writes to serial interface or TX buffer
SERSTAT	Reads status of serial interface
I2CPORT	Definition and import of I^2C bus interface
I2CSTAT	Reads status (ready/busy or present/absent) of I^2C bus interface
I2CINP	Reads from I^2C bus interface
I2COUT	Writes to I^2C bus interface
EEPROM	Read/write EEPROM

```
Program Test;                {Program Name}

Device . . .                 {Hardware Declaration}

Import . . .                 {Import System Functions}

Define . . .                 {Hardware Definition}

Implementation               {Program Start}

Const . . .                  {Declaration of Constants}

Var . . .                    {Declaration of Variables}

Procedure ABC;               {Procedure Header}
begin
  . . .
end;

Function CDE : boolean;      {Function Header}
begin
  . . .
  Return(a > b);             {Result of Function}
end;

begin                        {Main Program}
  EnableInts;
  Loop
    . . .
    ABC;
    x:= CDE;
  EndLoop;
end.
```

Listing 4-2
Structure of an AVR Pascal source text.

4.5 AVR BASIC from Silicon Studio

Silicon Studio developed AVR BASIC to support easy programming of AVR microcontrollers in a high-level language. Figure 4-25 shows the AVR BASIC programming environment.

Any program written in AVRASM can be rewritten for AVR BASIC to produce exactly the same code (on the same target). To provide an easy translation path (from any assembly source code), AVR BASIC supports many Assembly-like tokens directly. AVR BASIC ASM tokens do a little more than AVRASM. Instruction and operand combinations that are not available as single AVR instructions are still translated. For example;

```
Eor R0,   %00001111
```

is valid for AVR BASIC, translated to two instructions (`ldi` and `eor`).

For a detailed description of the use of AVRASM versus AVR BASIC, please check Assembler and BASIC source files on the Web site of Silicon Studio. The URL of Silicon Studio is listed in the Appendix. Those files include all AVRASM instruction variants with corresponding BASIC sources producing the same AVR code.

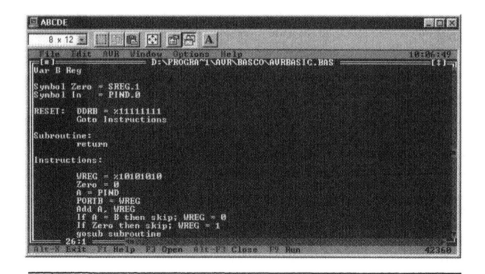

Figure 4-25
AVR BASIC programming environment.

Programs translated to AVR BASIC using the ASM style syntax are not as compact (in terms of source code) and readable as programs fully rewritten for AVR BASIC.

Table 4-11 show the relation between AVR Assembler and AVR BASIC instructions.

Table 4-11
Conversion Table from AVR Assembler to AVR BASIC.

AVR Assembler	AVR BASIC (AT90S1200)
BSET, BCLR, SBI, CBI, BST, BLD, SEC, CLC, SEN, CLN, SEZ, CLZ, SEI, CLI, SES, CLS, SEV, CLV, SET, CLT, SEH, CLH, CLR, SET MOV, LDI, LD, ST, IN, OUT, SBR,[1] CBR[1]	Dest = Source
ADD	add
ADC	adc
AND	and
ASR	asr
BRBS, BRBC, BREQ, BRNE, BRCS, BRCC, BRSH, BRLO, BRMI, BRPL, BRGE, BRLT, BRHS, BRHC, BRTS, BRTC, BRVS, BRVC, BRIE, BRID	If BitVar Then Label
COM	com
CPSE	If Dest=Source Then Skip
CP, CPI	cp
DEC	dec
EOR	eor
INC	inc
NEG	neg
LSL	lsl
LSR	lsr
OR, ORI	or
RCALL	gosub
RET	return
RETI	reti
ROL	rol
ROR	ror
TST[2] (See CBR)	and Source, Source
SBIC, SBIS, SBRC, SBRS	If BitVar Then Skip
SLEEP	sleep
SUB, SUBI	sub
SBC, SBCI	sbc
SWAP	swap
WDR	wdr

[1] CBR/SBR are not AVR core instructions. Use and/or commands in BASIC to clr/set multiply bits or BitVar=Const to clr/set one single bit.

[2] TST is not an AVR core instruction; it is translated to AND by AVRASM.

An AVR BASIC program example is described in the application portion of this book.

More examples for programming in AVR BASIC can be found on the related Web site from Silicon Studio. At the URL http://www.avrbasic/com, you will find a growing number of application notes and sample projects ready for download. Figure 4-26 shows the contents of this Web site as of October 1997.

4.6 Programmer and Evaluation Boards

Programmer and evaluation boards are described in an unique section here. All microcontrollers of the AVR family are in-circuit progammable, so a few connections between the microcontroller to be programmed and the programming device are sufficient.

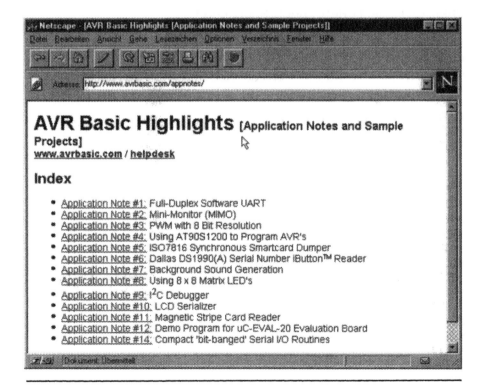

Figure 4-26
AVR BASIC Application Notes and Sample Projects (10/19/97)

AVR RISC Microcontroller Handbook

The market for programmers and evaluation boards grew rapidly with the arrival of the AT90S1200—the first member of the AVR microcontroller family.

4.6.1 AVR Development Board from Atmel

Atmel designed the AVR Development Board to help new AVR users get quickly acquainted with the AVR microcontroller devices.

The AVR Development Board can be used as a breadboard for testing new designs before PCB development, and also as a programmer. Figure 4-27 shows Atmel's AVR Development Board assembled with an AT90S1200 microcontroller inside the 40-pin socket for an AT90S8515.

Figure 4-27
AVR Development board from Atmel.

The AVR Development Board has the following features:

• Pushbuttons connected to PortB and LEDs connected to PortD for general use
• All ports accessible through header connectors
• Serial port with RS-232 conform signal levels for general use
• Serial programming (over a second serial port)

The AVR Development Board works with the AVR Programming Software AvrProg running under Windows95/NT. Outside these environments, you can get almost the same functionality with some DOS programs.

AvrProg can be started from AVR Studio from menu **Tools > Program Device** or directly from Windows. The programming software detects a connected and powered AVR Development Board on any COM port of the PC automatically.

Figures 4-28 and 4-29 show the two windows for programming and setup with the AvrProg programming software.

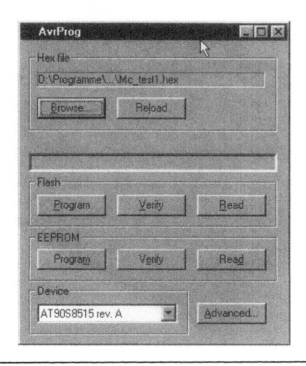

Figure 4-28
AvrProg programming window.

AVR RISC Microcontroller Handbook

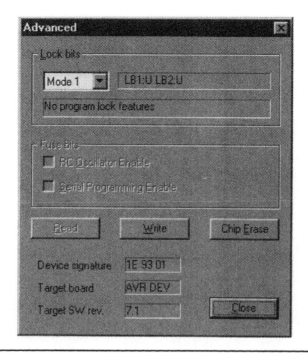

Figure 4-29
AvrProg advanced functions.

4.6.2 ISP Starter Kit from Equinox

Equinox Technologies supports code development for the AVR family with its ISP Starter Kit. This kit consists of the following parts:

• Micro-ISP serial download programmer
• AVR evaluation module
• Atmel AT90S1200 DIL microcontroller
• Micro-ISP for Windows software
• AVR Assembler and Simulator software
• Atmel 90S series data sheets
• "In-System Programming" application note
• Atmel CD-ROM data book

Figure 4-30 shows the hardware parts of the AVR Starter Kit. On the left side, you can see the AVR Evaluation Module uC-EVAL-20 (for 20-pin AVR devices). On the right side is the Micro-ISP serial download programmer.

Figure 4-30
Hardware of AVR Starter Kit from Equinox.

The AVR Evaluation Module uC-EVAL-20 must be connected to an unused LPT port of a PC and must be powered from the target or by an external power supply. The evaluation module is assembled with an AT90S1200.

To support immediate code development, the ISP Starter Kit also contains Atmel's AVR Assembler and Simulator. After the code, is assembled, it can be tested with the Simulator. To program the target microcontroller in-system, without removing it from the target socket, use the Micro-ISP serial download programmer.

The Micro-ISP serial download programmer hardware is controlled by Windows program. Figure 4-31 shows the opening window of the Micro-ISP Programmer software.

Figure 4-32 shows the memory contents for an AT90S1200 project decribed later. It can be seen that not only code memory (flash) is loaded. In the EEP-ROM, ASCII data are stored.

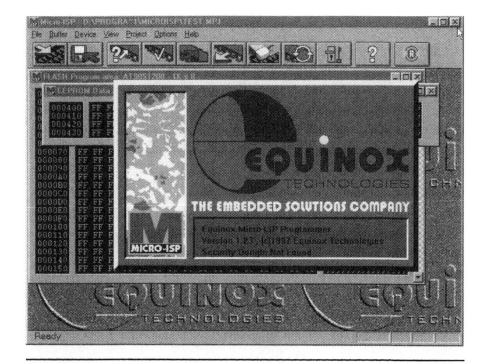

Figure 4-31
Micro-ISP Programmer opening window.

4.6.3 SIMMSTICK from Silicon Studio

SiliconStudio and DonTronics developed the SIMMSTICK family of multi-function microcontroller modules. These modules have different sizes but an unique SIMM30 edge connector.

The DT104 board shown in Figure 4-33 can be used as both a programmer board and a target board for 20-pin AVR microcontrollers (DIP20).

The circuitry on the board consists of an AVR (DIP20) microcontroller, a voltage regulator (TO-92), I²C serial EEPROM (24CXX) or I²C Serial RAM, an oscillator circuit, reset/brownout circuitry, I/O drivers, pull-up resistors, and an optional RS-485 driver.

For the 40-pin AVR microcontroller (DIP40), the DT103 board shown in Figure 4-34 was developed. It can be used as both a programmer board and a target board.

The circuitry on the board consists of an AVR (DIP40) microcontroller, two pushbuttons, two LEDs, a voltage regulator (TO-92), SPI or I²C serial EEP-ROM, an oscillator circuit, reset circuitry, and an LCD interface. The LCD header pins suit both the 14- and 16-pin types.

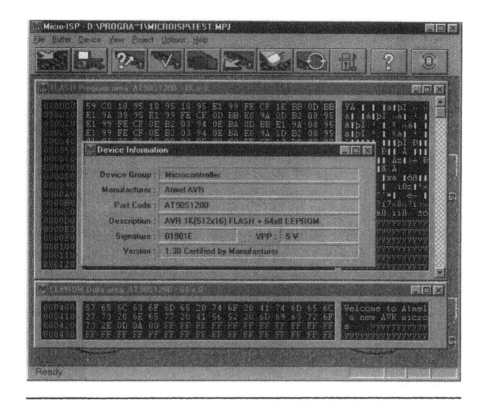

Figure 4-32
Display of the contents of flash memory and EEPROM for AT90S1200.

Figure 4-33
SIMMSTICK Board for AT90S1200.

Figure 4-34
SIMMSTICK board for AT90S8515.

For both SIMMSTICK boards the needed DOS programmer software PIP04 can be downloaded from Silicon Studio's Web site.

4.6.4 Parallel Port Programmer BA1FB

Jerry Meng's PC-based AVR programmer BA1FB is an easy-to-build, beginner's development programmer. It can also be connected to the application system via CLK, MOSI, MISO, and GND for in-circuit serial programming over SPI.

No power supply is needed. The programmer supports only Intel HEX file format. Because SPI program mode does not support RC OSC fuse mode, users should order AT90S1200 for the crystal OSC and AT90S1200A for the internal RC OSC.

Figure 4-35 shows the schematic of the parallel Port programmer BA1FB. Note that there are no external components. The AVR microcontroller to be programmed will be driven entirely from the PC's parallel port. The programmer software is freeware. You can download the whole project from the author's Web site, for example.

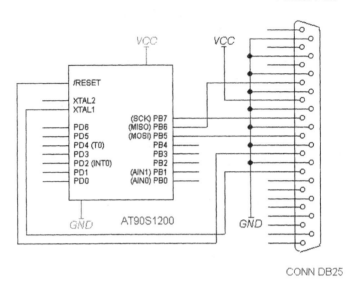

CONN DB25

Figure 4-35
Parallel port programmer BA1FB.

4.6.5 Serial Port Programmer PonyProg

If you prefer the serial port for external equipment on the PC then the serial port programmer PonyProg developed by Claudio Lanconelli could be of interest. Figure 4-36 shows the schematics of the PonyProg programmer. A PCB is also available.

At this time the programmer is still under development. Contact Claudio's Web site for actual hard- and software. You will find Claudio on the Web at http://www.cs.unibo.it/~lancone.

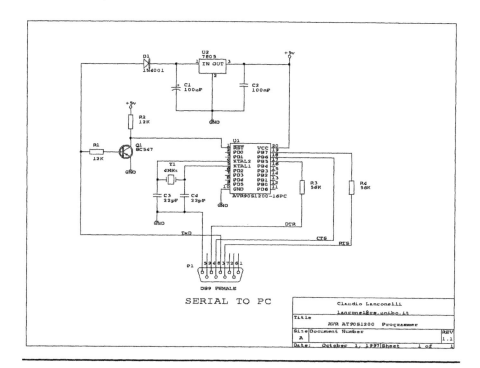

Figure 4-36
Serial port programmer PonyProg.

Example Programs 5

To help provide the knowledge needed for programming one's first projects with a new microcontroller, some program examples are always helpful.

For AVR microcontrollers, we have come to know Atmel's AVR Assembler, IAR's C compiler, and the AVR BASIC from Silicon Studio in the last chapter. Example programs explained here will give a first impression of the use of these interesting microcontrollers.

Starting with some examples programmed in Assembler should give an idea of the use of the new Assembler instructions and directives. Because of their different hardware resources (stack and peripherals), we separate the example programs for the AT90S1200 and the AT90S8515 microcontroller.

Some later C examples will show the use of a high-level language for programming microcontrollers with limited resources. The C compiler does not support the AT90S1200 microcontroller with its hardware stack, so all example programs refer to the AT90S8515 microcontroller.

The AVR BASIC examples will give an impression of further alternative programming environments.

5.1 Example Programs in AVR Assembler

The following sections introduce some example programs written in AVR Assembler for AT90S1200 and AT90S8515. All were tested on Atmel's evaluation board.

After an adaption of the interrupt vector table and stack pointer initialization, all programs written for the AT90S1200 work on the AT90S8515, AT90S2313, or AT90S4414. The programs written for the AT90S8515 use in most cases the enhanced internal hardware of this type of microcontroller and cannot work on the AT90S1200 because of its reduced resources. Before these programs are used on the AT90S2313 or AT90S4414, the availability of the required resources must first be checked.

5.1.1 Assembler Programs for the AT90S1200

First Test with the AT90S1200 For those who are unexperienced with a new microcontroller and its new programming environment, a very simple test program helps in testing the single steps from writing the source code, debugging and simulation, to programming and testing the microcontroller itself.

For the first test, we take an AT90S1200 with a minimum of external elements. Figure 5-1 shows the circuit diagram. The circuitry used is part of Atmel's development board for AVR microcontrollers.

The reset and clock circuitry is unexceptional. LEDs in series with a current limiting resistor are connected to each pin of PortB. Pin2 of PortD serves as input where a simple key with a pull-up resistor is connected.

The structogram shown in Figure 5-2 explains the function of this simple I/O test program. After some initialization, the program queries the key. The program waits until the key is pressed. After the pressed key is detected, the bit pattern initialized after reset is changed and output to the LED afterwards. Now the program waits until the key is released, and repeats the whole process endlessly.

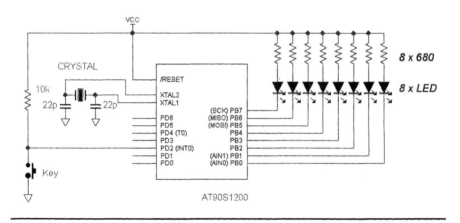

Figure 5-1
Test circuitry.

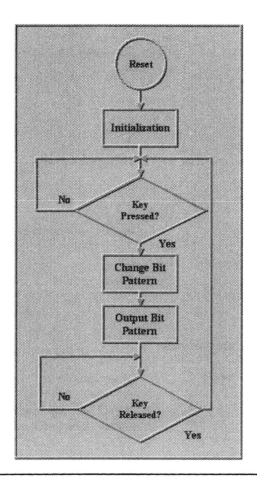

Figure 5-2
Structogram of I/O test program.

Listing 5-1 shows the program source in detail.

After a descriptive header, the program source starts with some directives. The first directive defines which device to assemble for. We choose the AT90S1200 for this first experiment. To use all I/O register names and I/O register bit names appearing in the data book and this description the file 1200def.inc is included. The list and nolist directives are described in Chapter 4. In the subsequent directive, the variable temp is defined and assigned to register R16.

```
;********************************************************
;* File Name              :mc_test0.asm
;* Title                  :First Microcontroller Test
;* Date                   :08/31/97
;* Version                :1.0
;* Target MCU             :AT90S1200
;*
;* DESCRIPTION
;* Test of simple port I/O
;********************************************************


;***** Directives

.device at90s1200           ; Device is AT90S1200
.nolist
.include "1200def.inc"
.list


.def    temp     = r16

;***** Interrupt vector table

        rjmp    RESET       ; Reset handle
        reti                ; External Interrupt0 handle
        reti                ; Overflow0 Interrupt handle
        reti                ; Analog Comparator Interrupt handle

;***** Main

RESET:  ser     temp
        out     DDRB,temp   ; PORTB = all outputs
        sec
        ldi     temp, $FE
        out     PORTB, temp ; Set bit pattern

loop:   sbic    PIND,2      ; Wait until key is pressed
        rjmp    loop
        rol     temp        ; Rotate bit pattern
        out     PORTB, temp ; Output rotated bit pattern
wait:   sbis    PIND,2      ; Wait until key is unpressed
        rjmp    wait
        rjmp    loop        ; Repeat forever
```

Listing 5-1
Test program mc_test0.asm.

This simple I/O test program operates without interrupts. Including an interrupt vector table anyway is a matter of taste. Further program enhancements are thus prepared for, and the program structure always remains the same.

The reset vector points directly to the initialization part. PortB is initialized as output and PortD serves as input by default conditions after reset. Afterwards, the bit pattern is set as shown in Figure 5-3.

The changes in bit pattern result from rotation through the Carry flag. To generate a rotation of a bright LED, the variable temp is initialized to FEH and the Carry flag set. Figure 5-3 shows the rotating bit pattern (a "0" switches the LED on).

In an endless loop, the steps explained in Figure 5-2 are executed. The bit pattern also changes by a bit rotation through the Carry flag as shown in Figure 5-3.

If you test this program with a simple key, in some cases you will get an LED jumping over more than one position. This is correct behavior because each key is bouncing. If key bouncing is a problem in an application, you will have to debounce the keys by hardware or software means.

Including Interrupts   If the environment works, it is more interesting to have a look at handling the resources of the microcontroller. In microcontroller applications, interrupts make possible time-critical responses to peripheral events.

To test some interrupt features, Pin2 of PortD now serves as input for the external interrupt INT0. The circuitry, for this program example is the same as shown in Figure 5-1. The bit pattern will be output on PortB periodically, and a pressed key must change the refreshing period for this output.

In the related program we will find three independent branches. Figure 5-4 shows the structogram of this test program.

Nearly all the work will be done by two independent interrupt handlers. The interrupt handler for timer interrupt TOV0 will be activated on timer overflow. It sends the bit pattern to PortB for switching the LEDs, followed

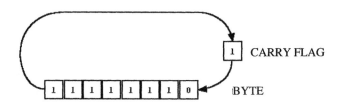

Figure 5-3
Rotating the bit pattern through Carry.

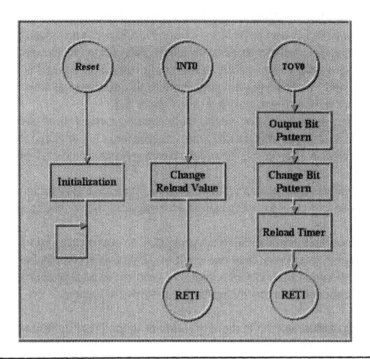

Figure 5-4
Structogram of microcontroller test program.

by a change of this bit pattern and a reload of the timer for the next timer period. The interrupt handler for the external interrupt INT0 requested by the pressed key changes the reload value only. After changing the reload value, the blink frequency of LED connected to PortB will change. Now, the main program has to initialize the microcontroller's hardware and some variables and can then go in an endless loop.

Listing 5-2 shows the complete microcontroller test program. Again you will find the same structure.

The structure of the source text in Listing 5-2 is the same as in Listing 5-1.

The directive .nolist before including the file 1200def.inc avoids the listing of this included file in the list file. After the include, the list directive is notified for listing the rest of the source. If there are macros in the program source, the listmac directive list the macro sequence in the list file, as you can see in Listing 5-3 at the end of this section.

In the next directives, three variable names are defined and assigned to registers. The variable temp locates in R16, byte in R17, and reload in R18.

AVR RISC Microcontroller Handbook

```
;*******************************************************
;* File Name          :mc_test1.asm
;* Title              :Microcontroller Test
;* Date               :28/03/97
;* Version            :1.0
;* Target MCU         :AT90S1200
;*
;* DESCRIPTION
;* Test of Timer0 Overflow and External Interrupt on AVR Evaluation Board
;* Oscillator frequency CK is 4 MHz (T = 0.25 us)
;*******************************************************

;***** Directives

.device at90s1200            ; Device is AT90S1200
.nolist
.include "1200def.inc"
.list
.listmac

.def    temp  = r16
.def    byte  = r17
.def    reload = r18

;***** Macros

.MACRO  SBIs_HR              ; Set Bits in I/O Register 32 up
        in     temp,@0
        ori    temp, @1
        out    @0, temp
.ENDMACRO

;***** Interrupt vector table
        rjmp   RESET         ; Reset handle
        rjmp   EX_INT0       ; External Interrupt0 handle
        rjmp   OVF0          ; Overflow0 Interrupt handle
        reti                 ; Analog Comparator Interrupt handle

;***** Interrupt handlers
```

Listing 5-2
Test program mc_test1.asm.   (*Continued*)

```
EX_INT0 swap    reload          ; Rotate Reload
        reti


OVF0:   out     PORTB, byte     ; Output Bit Pattern
        rol     byte            ; Rotate Bit Pattern
        out     TCNT0, reload   ; Reload Timer/Counter0
        reti


;***** Main


RESET:  ldi     reload,$07      ; Initialize Reload Value


        sec
        ldi     byte, $FE
        out     PORTB,byte      ; Initialize PortB


        ser     temp
        out     DDRB,temp       ; PORTB = all outputs


        ldi     temp, 0b00000101
        out     TCCR0, temp     ; Prescaler CK/1024 => 2,56 ms


        ldi     temp, TOV0<<1
        out     TIMSK, temp     ; T/C0 Interrupt Enable


        SBIs_HR MCUCR,0b00000010; Interrupt on falling edge of INT0


        SBIs_HR GIMSK,0b01000000; External Interrupt0 Enable


        sei                     ; Global Interrupt Enable

loop:   rjmp    loop            ; Repeat forever
```

Listing 5-2
Continued

In the next section a macro is defined. Because the SBI instruction sets bits
in the lower I/O registers, only the macro SBIs_HR is defined. SBIs_HR
masks the content of an I/O register with an immediate bit pattern. A "1" in
this bit pattern forces the related bit to "1," while a "0" lets the related bit remain unchanged.

```
;**********************************************************************
;* File Name         :mc_test1.asm
;* Title             :Microcontroller Test
;* Date              :28/03/97
;* Version           :1.0
;* Target MCU        :AT90S1200
;*
;* DESCRIPTION
;* Test of Timer0 Overflow and External Interrupt on AVR Evaluation Board
;* Oscillator frequency CK is 4 MHz (T = 0.25 us)
;**********************************************************************

;***** Directives

.device at90s1200              ; Device is AT90S1200
.nolist
.listmac

.def    temp      = r16
.def    byte      = r17
.def    reload    = r18

;***** Macros

.MACRO   SBIs_HR               ; Set Bits in I/O Register 32 up
         in       temp,@0
         ori      temp, @1
         out      @0, temp
.ENDMACRO

;***** Interrupt vector table

000000 c009   rjmp   RESET      ; Reset handle
000001 c002   rjmp   EX_INT0    ; External Interrupt0 handle
000002 c003   rjmp   OVF0       ; Overflow0 Interrupt handle
000003 9518   reti              ; Analog Comparator Interrupt handle

;***** Interrupt handlers
```

Listing 5-3
List file mc_test1.lst.   (*Continued*)

```
000004 9522   EX_INTO:swap reload ; Rotate Reload
000005 9518     reti


000006 bb18   OVF0:  out    PORTB, byte ; Output Bit Pattern
000007 1f11     rol  byte            ; Rotate Bit Pattern
000008 bf22     out  TCNT0, reload ; Reload Timer/Counter0
000009 9518     reti


;***** Main


00000a e027   RESET: ldi   reload,$07 ; Initialize Reload Value


00000b 9408   sec
00000c ef1e   ldi   byte, $FE
00000d bb18   out   PORTB,byte   ; Initialize PortB


00000e ef0f   ser   temp
00000f bb07   out   DDRB,temp    ; PORTB = all outputs


000010 e005   ldi   temp, 0b00000101
000011 bf03   out   TCCR0, temp ; Prescaler CK/1024 => 256 us


000012 e002   ldi   temp, TOV0<<1
000013 bf09   out   TIMSK, temp ; T/C0 Interrupt Enable


000014 +      SBIs_HR MCUCR,0b00000010 ; Interrupt on falling edge of INT0
000014 b705   in    temp,mcucr
000015 6002   ori   temp, 0x2
000016 bf05   out   mcucr, temp
         .ENDMACRO


000017 +      SBIs_HR  GIMSK,0b01000000 ; External Interrupt0 Enable
000017 b70b   in    temp,gimsk
000018 6400   ori   temp, 0x40
000019 bf0b   out   gimsk, temp
         .ENDMACRO


00001a 9478   sei                    ; Global Interrupt Enable
00001b cfff   loop: rjmp  loop  ; Repeat forever


Assembly complete with no errors.
```

Listing 5-3
Continued

In the next section we find the interrupt vector table. Beside the reset vector, we have three interrupts in AT90S1200. The external interrupt and the timer overflow interrupt are used in this example; therefore, a relative jump is notified in the interrupt vector table. On the location for the unused analog comparator interrupt, a return from interrupt is placed for all cases.

Before we come to the main part of the program source, the section for interrupt handlers follows. The handlers for the external interrupt requested by pressing the key at pin INT0 swaps the reload value. Swap means a change from high-nibble and low-nibble of that byte. The handler for the timer overflow interrupt writes the value of the variable `byte` to PortB for LED display rotates this bit pattern to prepare a changed display and reloads the timer register for the next timer period.

In the main part of the program source follow many initializations before the program is set free for execution. At first the reload value is initialized to 7. Because of the swapping in the external interrupt handler, the timer will work with 7 and 70_H as reload values. The resulting bit pattern refresh rate is 64 ms for reload value 7 and 37 ms for reload value 70_H.

In the next lines the bit pattern in variable `byte` is initialized. The changes in bit pattern result from rotation as described in the previous section.

In the next steps, the initialization of PortB as output and the setup of the prescaler for CK/1024 follow. The resulting internal counter clock is 1024/4MHz = 256 ms, so a maximum timer period of 65.536 ms can be achieved. After the timer overflow interrupt enable bit is set, the timer is prepared for further operation.

The external interrupt can be requested by different events on pin INT0. In this example, a falling edge on pin INT0 should request the interrupt. By setting the registers MCUCR and GIMSK, the external interrupt is initialized and enabled. The INT0 pin is selected as input by reset default condition; therefore, an initilization of the data direction register DDRD is not needed here.

The last instruction before entering the endless loop is enabling the Global Interrupt. After this Global Interrupt Enable, both prepared interrupt handlers are activated and waiting for requesting events.

Closing this section, the generated list file shows the results of assembling the program source.

Programming Timer Periods and Duty Cycles

In microcontroller applications, timers generate periodic events, often to start an analog-to-digital conversion, output messages, query another networked microcontroller node, and so forth. In the preceding section, we used Timer/Counter0 to generate periodic interrupt for data output to LEDs.

Table 5-1 shows the maximum timer period and timer resolution for a microcontroller clock frequency of 4 MHz. The prescaler determines the

Table 5-1
Maximum timer period and resulting resolution for T/C0 at 4-MHz clock.

	T/C0 Clock [Hz]	Max. Timer Period [ms]	Timer Resolution [µs]
CK	4,000,000.000	0.064	0.250
CK/8	500,000.000	0.512	2.000
CK/64	62,500.000	4.096	16.000
CK/256	15,625.000	16.384	64.000
CK/1024	3,906.250	65.536	256.000

maximum timer period. The timer resolution results from 8-bit register TCNT0 to 1/256 of this period.

To generate a square wave with a defined frequency, a prescaler and a reload value must be calculated. On a square wave with a frequency of 1 kHz, each 500 µs the output pin must be toggled. Figure 5-5 shows the square wave to be generated.

Looking in Table 5-1, we find the prescaler CK/8 with a maximum timer period of 512 µs. The resolution of time is 2 µs, and the timer must count 250 states before overflow. The reload value for the up-counter must be, therefore, $256 - 250 = 6$.

Listing 5-4 shows the program source, which differs slightly from Listing 5-2. Only the timer overflow interrupt is working. The other lines in the interrupt vector table, therefore, contain the return from interrupt instruction.

A square-wave output serves Pin0 of PortB (PB0). The timer overflow interrupt handler must now toggle this pin by XORing the read value with a mask before outputting the new calculated value. In the main part of the program source, some initialization takes place before the program runs in an endless loop.

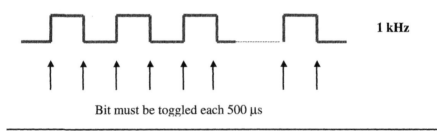

Bit must be toggled each 500 µs

Figure 5-5
Square-wave generation.

AVR RISC Microcontroller Handbook

```
;**********************************************************
;* File Name                :timer0.asm
;* Title                    :Square Wave output
;* Date                     :23/08/97
;* Version                  :1.0
;* Target MCU               :AT90S1200
;*
;* DESCRIPTION
;* Square Wave Generation of 1 kHz by Timer0 Overflow
;**********************************************************

;***** Directives

.device at90s1200                ; Device is AT90S1200
.nolist
.include "1200def.inc"
.list
.listmac

.def    temp      = r16
.def    mask      = r17
.def    reload    = r18

;***** Interrupt vector table

        rjmp    RESET            ; Reset handle
        reti                     ; External Interrupt0 handle
        rjmp    OVF0             ; Overflow0 Interrupt handle
        reti                     ; Analog Comparator Interrupt handle

;***** Interrupt handlers

OVF0:   in      temp, PORTB      ; Input Bit Pattern
        eor     temp, mask       ; Toggle Bit0
        out     PORTB, temp      ; Output PB0
        out     TCNT0, reload    ; Reload Timer/Counter0
        reti

;***** Main

RESET:  ldi     mask, $01        ; Initialize Mask Byte
        ldi     reload,$06       ; Initialize Reload Value
        sbi     DDRB,PB0         ; PB0 = output
        ldi     temp, 0b00000010
        out     TCCR0, temp      ; Prescaler CK/8 => 2 us
        ldi     temp, TOV0<<1
        out     TIMSK, temp      ; T/C0 Interrupt Enable
        sei                      ; Global Interrupt Enable
loop:   rjmp    loop             ; Repeat forever
```

Listing 5-4
Square-wave generation.

For a practical test of the generated timing, we can connect digital time-measurement equipment or an oscilloscope to PB0. Time adjustments in steps of 2 μs are possible by changing the reload value for the timer. To detect the bit toggling by an LED, the prescaler must increase. Changing the prescaler from CK/8 to CK/1024 changes the frequency from 1 kHz to 7.8 Hz—easily visible by an LED.

Changing the reload value for the timer gives us a variation of the timer period on the fly. The generation of pulse-width-modulated signals uses periodic switching between two dependent timer reload values.

Before we start with the program example, Figure 5-6 explains the principle of a pulse-width modulated signal with four states, that is, 2 bits. The four states differ in the ratio between Hi and Lo time of the signal. In more technical terms, the duty cycle varies from 0, ¼, ½, to ¾. If we integrate over the whole period, we find the same values again.

The pulse signals shown in Figure 5-6 can be generated by the microcontroller's timer and output to any pin. The Timer/Counter0 of the AT90S1200 is an 8-Bit timer/counter and makes possible 256 different duty cycles. The resulting wave form output on any pin can be integrated by a resistor and a capacitor to provide an analogous voltage.

To see the results, first look at Table 5-2 and Figure 5-7. Both list the voltage generated by pulse-width modulation and measured with a digital voltmeter. One single step is about 19 mV. The overall linearity looks very good.

For state 0, we cannot measure zero volts because the voltage on any pin in Lo state does not go to zero. The resulting drain–source voltage depends on the current flowing through that transistor. The data sheet specifies a maxi-

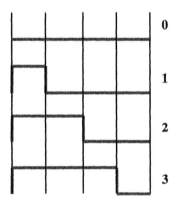

Figure 5-6
PWM Principle.

Table 5-2
Measured voltage of PWM with Timer/Counter0.

Byte	$00	$10	$20	$30	$40	$50	$60	$70	$80	$90	$A0	$B0	$C0	$D0	$E0	$F0	$FF
Voltage	.073	.393	.696	.99	1.29	1.59	1.895	2.20	2.50	2.80	3.10	3.40	3.71	4.01	4.31	4.61	4.89

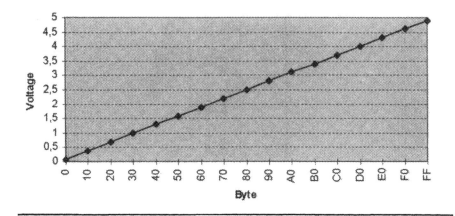

Figure 5-7
Voltage output by pulse-width modulation.

mum remaining voltage of 0.5 V at 20 mA, that is, a resistance of less than 25 Ω from pin to ground.

Directing our attention to the program example itself, we start with the structogram. Figure 5-8 shows the structogram of the timer overflow interrupt handler only because all activity outside of the initialization is done by this handler.

A pulse-width-modulated signal normally consists of two signal phases—a Hi-phase and a Lo-phase. One exception is the state zero where no Hi-phase exists. By testing the reload value for zero and setting the flag, a possible Hi-phase is avoided. The next Hi-phase will first be possible when this reload value is changed from zero to another valid value.

Because of these two phases, the timer overflow interrupt handler must decide which phase must be handled. A flag signals which phase must be handled next. In each phase there are three activities:

(1) Swiching the pin Hi or Lo
(2) Reloading the timer with the Hi-value (byte) or Lo-value (256-byte)
(3) Clear or set the flag for marking the phase

The program shown in Listing 5-5 is an implementation example for the generation of a pulse-width-modulated signal according to Figure 5-8.

During initialization, the variable byte is loaded with 80_H. This value serves as the reload value for the timer, and a symmetrical wave form results. To change the duty cycle, change the instruction written in italics, reassemble the program source, and download the changed code. This is how the measurements of characteristics (Table 5-2 and Figure 5-7) were generated.

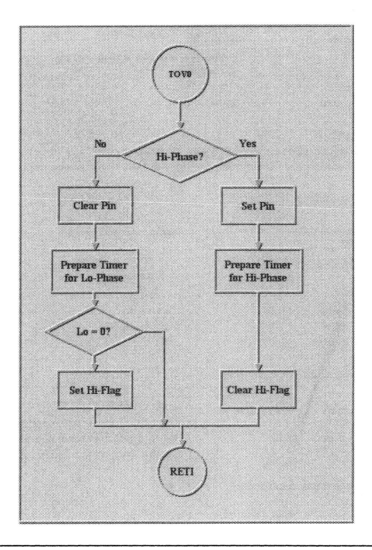

Figure 5-8
Interrupt handler for pulse-width modulation.

Software Delay In some cases no internal timer can be used to generate time delays. Software loops with included `nop` instructions or without any instructions can help to kill time. The drawback of this method is that the microcontroller is busy although no application function is executed.

The time delay must be calculated by the number of instruction cycles times the execution time for one instruction or clock cycle. If the delay loop is interrupted, then the execution time for the interrupt handler must be added. In interrupt-controlled applications, therefore, no exact time prediction is possible.

```
;********************************************************************
;* File Name            :pwm0.asm
;* Title                :Pulse Width Modulation
;* Date                 :23/08/97
;* Version              :1.0
;* Target MCU           :AT90S1200
;*
;* DESCRIPTION
;* Pulse width modulation by Timer/Counter0
;********************************************************************

;***** Directives

.device at90s1200                ; Device is AT90S1200
.nolist
.include "1200def.inc"
.list
.listmac
.def    temp    = r16
.def    byte    = r17
.def    reload  = r18

;***** Interrupt vector table

        rjmp    RESET           ; Reset handle
        reti                    ; External Interrupt0 handle
        rjmp    OVF0            ; Overflow0 Interrupt handle
        reti                    ; Analog Comparator Interrupt handle

;***** Interrupt handlers

OVF0:   brts    high            ; Test T Flag
low:    cbi     PORTB, PB0      ; Clear PB0
        mov     reload, byte
        out     TCNT0, reload   ; Reload Timer/Counter0
        tst     byte            ; Test Reload Value for Zero
        breq    low1
        set                     ; Set T Flag if Reload # Zero
low1:   reti
```

Listing 5-5
Generation of pulse-width-modulated output (pwm0.asm).

```
high:    sbi    PORTB, PB0      ; Set PB0
         neg    reload
         out    TCNT0, reload   ; Reload Timer/Counter0
         clt                    ; Clear T Flag
         reti


;***** Main

RESET:   ldi    byte, $80
         mov    reload,byte     ; Initialize Reload Value

         sbi    DDRB,PB0        ; PB0 = output

         ldi    temp, 0b00000010
         out    TCCR0, temp     ; Prescaler CK/8 => 2 us

         ldi    temp, TOV0<<1
         out    TIMSK, temp     ; T/C0 Interrupt Enable

         sei                    ; Global Interrupt Enable

loop:    rjmp   loop            ; Repeat forever
```

Listing 5-5
Continued

Because a simple loop allows only 256 runs through the loop, in general combined loops are used. Figure 5-9 shows a structogram of two combined loops.

If the inner loop is initialized with the value *n* and the outer loop with *m*, then *m* * *n* loop cycles are possible in all.

Listing 5-6 shows the implementation of a software-based time delay in the subroutine part. At the beginning of this subroutine, the variable delcnt is initialized with zero fix. The variable dly must be set before the function call.

In program delay.asm the variable dly is set to FF_H before the function call. The delay time is calculated by the following formula:

$$delay = \frac{1026 \cdot \text{dly} + 4}{f_{CLK}}$$

The differences in the formula *m**n come from instructions included in the loops.

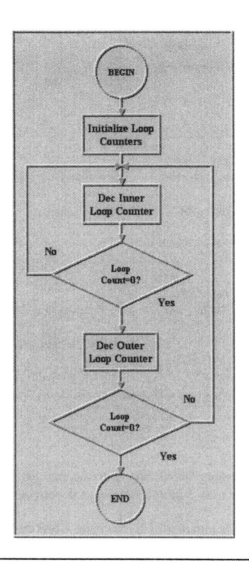

Figure 5-9
Software delay.

Under the setup conditions in Listing 5-6, we can calculate a delay of 65.4085 ms for an AVR microcontroller with a clock frequency of 4 MHz (such as Atmel's evaluation board). For longer delays, one or more loops must be connected together.

Query Keys Information about user activities is often signalized to the microcontroller by pressed keys. More or less complex keyboards are well

```
;******************************************************************
;* File Name            :delay.asm
;* Title                :Software Delay
;* Date                 :08/31/97
;* Version              :1.0
;* Target MCU           :AT90S1200
;*
;* DESCRIPTION
;* Software delay with counted loops
;******************************************************************

;***** Directives

.device at90s1200                ; Device is AT90S1200
.nolist
.include "1200def.inc"
.list

.def      temp    = r16
.def      dly     = r17
.def      delcnt  = r18          ; Loop Counter

;***** Interrupt vector table

          rjmp    RESET          ; Reset handle
          reti                   ; External Interrupt0 handle
          reti                   ; Overflow0 Interrupt handle
          reti                   ; Analog Comparator Interrupt handle

;***** Subroutines

DELAY:    clr     delcnt         ; Init Loop Counter
loop1:    dec     delcnt
          nop
          brne    loop1
          dec     dly
          brne    loop1
          ret
```

Listing 5-6
Software delay (delay.asm).   (*Continued*)

```
;***** Main

RESET:    ser    temp
          out    DDRB,temp      ; PORTB = all outputs
          out    PORTB, temp
          ldi    dly, $FF       ; Initialize delay of about 65 ms

loop:     cbi    PORTB, PB0

          ldi    dly, $FF       ; Initialize delay of about 65 ms
          rcall  DELAY          ; Number of cycles = 1026 * dly + 4

          sbi    PORTB, PB0

          ldi    dly, $FF       ; Initialize delay of about 65 ms
          rcall  DELAY          ; Number of cycles = 1026 * dly + 4

          rjmp   loop           ; Repeat forever
```

Listing 5-6
Continued

known in the world of personal computers. In microcontroller applications, a few keys will do.

On Atmel's AVR evaluation board, switches to GND with a pull-up resistor to VCC each are connected to PortD, while LEDs with current-limiting resistors in series are connected to PortB. If no key is pressed, all pins of PortD see Hi logic level. A pressed key will connect that pin to GND and Lo level will be detected on next query.

The program example shown in Listing 5-7 queries all pins of PortD in an endless loop and sends the result to PortB for display with the connected LEDs. To make the changes visible, a software delay was included.

The simple program in Listing 5-7 has one essential disadvantage. Independent of whether a key was pressed or not, the microcontroller queries the keys and outputs the result unchanged most times.

If the microcontroller must execute time-consuming routines, the time between two queries of the keys will increase. So it is possible that a key pressed only briefly will not be detected. In such cases, interrupt-controlled queries of pressed keys help to save the resources of a microcontroller.

In Figure 5-10, three keys are connected to PortD. The diodes build a logical AND for the key signals. Only if all keys are unpressed will PD2 see a logical Hi. Pressing any key will pull down the level on PD2. With its alternative function, this pull-down can generate an interrupt request.

```
;********************************************************************
;* File Name              :key0.asm
;* Title                  :Query a Key
;* Date                   :28/03/97
;* Version                :1.0
;* Target MCU             :AT90S1200
;*
;* DESCRIPTION
;* Query of keys in a loop
;********************************************************************

;***** Directives

.device at90s1200                ; Device is AT90S1200
.nolist
.include "1200def.inc"
.list

.def    temp     = r16
.def    dly      = r17
.def    delcnt   = r18           ; Loop Counter

;***** Interrupt vector table

        rjmp    RESET            ; Reset handle
        reti                     ; External Interrupt0 handle
        reti                     ; Overflow0 Interrupt handle
        reti                     ; Analog Comparator Interrupt handle

;***** Subroutines

DELAY:  clr     delcnt           ; Init Loop Counter
loop1:  dec     delcnt
        nop
        brne    loop1
        dec     dly
        brne    loop
        ret

;***** Main

RESET:  ser     temp
        out     DDRB,temp        ; PORTB = all outputs
        out     PORTB, temp      ; Set all outputs Hi
        ldi     dly, $80         ; Initialize delay
```

Listing 5-7
Polled query of keys (key0.asm).   (*Continued*)

```
loop:    in      temp, PIND
         out     PORTB, temp
         rcall   DELAY           ; Number of cycles = 1026 * del + 4
         nop
         rjmp    loop            ; Repeat forever
```

Listing 5-7
Continued

In Listing 5-8, a query of the keys takes place only after an external interrupt generated by the logical linked keys. The interrupt handle has the same function as the polling loop in Listing 5-7, but it is only executed if there was a pressed key.

Because of the limited number of I/O pins in most microcontrollers, keys are connected in matrixes (Figure 5-11). The internal pull ups avoid external resistors around the keypad.

Software UART Not all types of the AVR microcontrollers have an internal UART to support asynchronous serial communication. For communication between microcontrollers and peripherals, serial data interchange is of great interest because only a limited number of pins are involved.

To build an RS-232 interface for the AT90S1200 microcontroller, the only possible way is the implementation of a software UART.

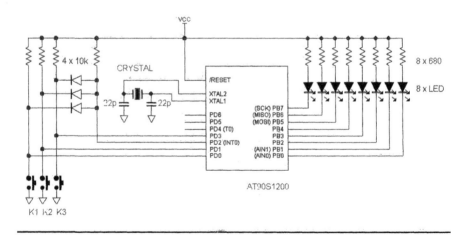

Figure 5-10
Interrupt-controlled query of keys.

```
;*********************************************************************
;* File Name             :key1.asm
;* Title                 :Query a Key
;* Date                  :28/03/97
;* Version               :1.0
;* Target MCU            :AT90S1200
;*
;* DESCRIPTION
;* Interrupt controlled query of keys
;*********************************************************************

;***** Directives

.device at90s1200                ; Device is AT90S1200
.nolist
.include "1200def.inc"
.list
.listmac

.def    temp    = r16
.def    byte    = r17

;***** Macros

.MACRO   SBIs_HR                 ; Set Bits in I/O Register 32 up
         in     temp,@0
         ori    temp, @1
         out    @0, temp
.ENDMACRO

;***** Interrupt vector table

         rjmp   RESET            ; Reset handle
         rjmp   EX_INT0          ; External Interrupt0 handle
         reti                    ; Overflow0 Interrupt handle
         reti                    ; Analog Comparator Interrupt handle

;***** Interrupt handlers

EX_INT0: in     temp, PIND       ; Read Pins of PortD
         out    PORTB, temp      ; Write Pattern to PortB
         reti
```

Listing 5-8
Interrupt-controlled query of keys (key1.asm).   (*Continued*)

```
;***** Main

RESET:    ser     temp
          out     DDRB,temp       ; PORTB = all outputs
          out     PORTB, temp     ; Set all outputs Hi


          SBIs_HR MCUCR,0b00000010; Interrupt on falling edge of INT0


          SBIs_HR GIMSK,0b01000000; External Interrupt0 Enable


          sei                     ; Global Interrupt Enable
loop:     rjmp    loop            ; Repeat forever
```

Listing 5-8
Continued

Robert Bush wrote the following software UART and has placed it in the public domain. Before we look at that implementation, some details that should help in better understanding the single steps of decoding the serial signal will be given.

The software UART described in this section operates with fixed parameters, such as 9600 bits per second (bps), one start bit, eight data bits, one stop bit, and no parity. For further descriptions of these parameters please look at the Appendix.

The bit stream shown in Figure 5-12 is seen directly from the microcontroller's pin. The line drivers normally used, such as MAX232 or equivalent, invert the polarity of this serial bit stream and work with higher voltage levels.

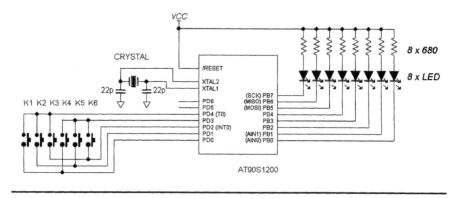

Figure 5-11
(2 × 3) key matrix.

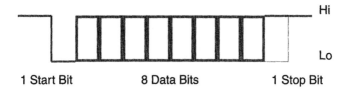

Figure 5-12
Bit stream of RS-232 signal.

In Figure 5-13, some points are marked. These points are important for sampling the serial signal and decoding it afterwards. The structogram shown in Figure 5-14 explains the single steps marked in Figure 5-13.

A falling edge of the sampled signal (0) marks an arriving start bit. After one-half of a bit time (1), the next sampling takes place to check if there is a valid start bit. If a Lo level is detected then the falling edge was the beginning of the start bit. Otherwise, the falling edge was only a faulty signal and all further sampling must be avoided until a valid start bit can be detected.

After a valid start bit, the serial input is expecting eight data bits. Before sampling the first data bit in a loop, the loop counter is initialized and the receive buffer is cleared.

The sampling cycle begins with a delay of one bit time, followed by sampling the data bit (2). Depending on the logic level of the received bit, a subroutine sets or clears the Carry flag and rotates the result in the Receive buffer. This cycle repeats eight times to sample all further data bits (3, 4, 5, 6, 7, 8, 9).

The stop bit is not sampled. Waiting one bit time after sampling all data bits delays the return from the serial input subroutine into the middle of the stop bit.

Listing 5-9 shows the complete program source for the software UART. Robert Bush took into consideration the RS-232 data interchange with RS-232 drivers and without.

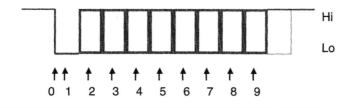

Figure 5-13
Sample points in bit stream.

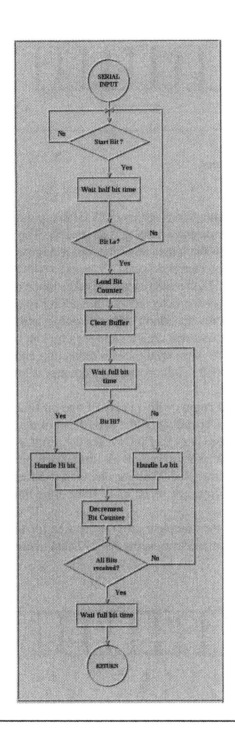

Figure 5-14
Sampling RS-232 bit stream.

```
        .include "1200def.inc"
;
; dsuart.asm
; Simple software async uart for AVR processors. Compatible with all AVR
; processors: AT90S1200, AT90S8515, etc.
; This version assumes that the soft uart pins are connected directly to a
; serial device (TTL level) or connected to an RS232 device through an RS232
; line driver (e.g., MAX232). Direct serial output to an RS232 device without
; using an RS232 line driver could be accomplished by reversing the bit logic
; instructions: e.g., all sbi/sbis instructions change to cbi/sbic
; and vice versa.
;
; Note: All labels,equates, and defines begin with 'd_.' This allows you
; to have multiple copies of the soft uart by simply doing a global replace
; of the 'd_' with character(s) of your own choice. Each copy of the soft
; uart occupies 52 words of program memory
;
; Author: R.Bush 8-26-97—PD release
;
        .org    0
        rjmp    start               ;execute a simple echo back test
;
; Primary routines:
; ---------
; d_sersetup:   Call to setup I/O bits for uart. Must call prior to using the
;               following routines.
; d_writeser:   Call with byte to send in d_txrx.
; d_readser:    Call to read serial port, byte returned in d_txrx.
;
; Registers used by software uart, define as needed (must be >= r16)
; ------------------------------------------------------------------
.def    d_bcntr = r23               ;bit counter
.def    d_srdel = r24               ;delay loop variable
.def    d_txrx  = r25               ;async serial data byte transmit/received
;
; Equates defining software uart, define as needed
; -------------------------------------------------
.equ    d_sport = portd             ;port supporting serial i/o
.equ    d_sin   = 0                 ;bit within port for serial data in
.equ    d_sout  = 1                 ;bit within port for serial data out
.equ    d_spin  = pind              ;port pins
.equ    d_sddr  = ddrd              ;data direction register for port
;
; Bit delays calculations:
; -----------------------
; where: baudrate = bps, fosc = crystal frequency in HZ.
;        note: round result to nearest whole number.
;
; overhead  = 20                    ;number of cycles to xmit a byte
```

Listing 5-9
Software UART (duart.asm).   (*Continued*)

```
; loopcount = 7                      ;number of cycles, primary delay loop
;
; oscrate   = 1 / fosc               ;xtal time period
; bitrate   = 1 / baudrate           ;baudrate bit time period
; looprate  = oscrate * loopcount    ;elapsed time per loop iteration
; adjust    = (overhead * oscrate) / looprate ;loop adjustment factor
;
; d_full    = (bitrate / looprate)—adjust ;delay loop count constant
;
; Note: d_full must be <= 255. Faster clocks and lower baudrates may necessitate
;       decreasing looprate by adding additional nop's to the primary delay
;       routine (d_bitdel). Each additional nop increases loopcount by one.
.equ      d_full      = 56           ;example for 4 MHz, 9600 baud
.equ      d_half      = 56 / 2
;
d_sersetup:    sbi      d_sport,d_sout   ;set output pin high
               sbi      d_sddr,d_sout    ;output pin data direction
               cbi      d_sddr,d_sin     ;insure input pin
               ret
;
; Software uart (serial port) for RS232 comm link.
; Assumes hardware pins used for software uart are connected
; to a MAX232 or similar driver.
; Receive serial data on i/o pin, place into d_txrx.
;
d_readser:     sbic     d_spin,d_sin     ;2/3 direct connect rs232 use
                                         ;sbic for MAX
               rjmp     d_readser        ;2 wait for start bit
               rcall    d_sdelay         ;3 delay for half bit time
;
               sbic     d_spin,d_sin     ;2/3 test for bit still there
               rjmp     d_readser        ;2
               ldi      d_bcntr,8        ;1 length of byte
               clr      d_txrx           ;1
;
d_recv:        rcall    d_bitdel         ;3 delay one full bit
;
               sbis     d_spin,d_sin     ;2/3 (skip) check for a '1' bit
               rjmp     d_rsz            ;2 handle a zero bit
               sec                       ;1
               ror      d_txrx           ;1
               rjmp     d_rsbin          ;2
;
d_rsz:         nop                       ;1
               clc                       ;1
               ror      d_txrx           ;1
               rjmp     d_rsbin          ;2
;
```

Listing 5-9
Continued

```
d_rsbin:          dec     d_bcntr           ;1
                  brne    d_recv            ;1/2
;                 com     d_txrx            ;compliment for direct RS232
                                            ;connection
                  rcall   d_bitdel          ;end up in middle of stop bit
                  ret                       ;4
;
; delay for 1/2 bit time
;
d_sdelay:         ldi     d_srdel,d_half    ;1 half bit delay
                  rjmp    d_bd1             ;2
;
; delay for full bit time — primary delay loop
;
d_bitdel:         ldi     d_srdel,d_full    ;1 full bit delay
                  rjmp    d_bd1             ;2 (used to balance loop timing)
d_bd1:            nop                       ;1
                  nop                       ;1
;
; If required, insert additional nop's here.
;
                  nop                       ;1
                  nop                       ;1
                  dec     d_srdel           ;1
                  brne    d_bd1             ;2/1
                  ret                       ;4
;
d_writeser:       ldi     d_bcntr,8         ;bit count
;                 com     d_txrx            ;compliment xmit byte for direct
                                            ;RS232 connect
                  cbi     d_sport,d_sout    ;set start bit
                  rcall   d_bitdel
d_wslp:           ror     d_txrx            ;1
                  brcs    d_xmt1            ;1/2
                  cbi     d_sport,d_sout    ;2
                  rjmp    d_xmt2            ;2
d_xmt1:           sbi     d_sport,d_sout    ;2
                  nop                       ;1
d_xmt2:           rcall   d_bitdel          ;delay full bit period
                  dec     d_bcntr           ;1
                  brne    d_wslp            ;1/2
                  sbi     d_sport,d_sout
                  rcall   d_bitdel          ;stop bit
                  ret
;
; Simple echo back test loop
;
```

Listing 5-9
Continued

(Continued)

```
start:
                rcall     d_sersetup      ;setup uart
test_loop:
                rcall     d_readser       ;read serial port
                inc       d_txrx
                rcall     d_writeser      ;echo back
                rjmp      test_loop       ;forever
;
; End
;
```

Listing 5-9
Continued

The notations in Listing 5-9 are valid on condition that (inverting) RS-232 drivers are used. In short-distance direct communication between microcontrollers and peripheral devices, there is no necessity for these drivers. The program duart.asm can be adapted to direct serial data interchange by reversing the bit logic instructions; for example, all sbi/sbis instructions change to cbi/sbic and vice versa. The byte stored in the Transmit/Receive buffer must complement as well. The required instructions were already noted in the receive and transmit subroutines, but are commented here.

Based on Robert Bush's program, an include file was written for an easy include in further applications. In the next section the use of that include file is demonstrated.

Using the external interrupt (INT0) for reception of serial data and the timer/counter (T/C0) to define the sample points is another way to implement an UART. Atmel's application note AVR304, "Half-Duplex Interrupt-Driven Software UART," describes the details of this approach.

An external interrupt is generated when a negative edge on the incoming serial signal is detected. The external interrupt handle disables further external interrupts and enables timer interrupts (bit-timer) because the UART must receive the incoming data.

The Timer/Counter0 Overflow Interrupt handle coordinates the transmission and reception of bits. This handle is automatically executed at a rate equal to the baud rate. When transmitting, this routine shifts the bit and sends it. When receiving, it samples the bit and shifts it into a buffer.

Tables in EEPROM   The AT90S1200 microcontroller does not understand the lpm instruction for access in code memory. In such cases, tables must be placed in the EEPROM.

In this program example, we place a text string into EEPROM for display purposes. An update of text for user dialogs is, thus, very easy. Listing 5-10 shows reading the EEPROM-based text string and sending it to RS-232.

To store the text string into EEPROM, we must initialize the required bytes with the .ESEG and .DB directives. To allow it to refer to the initalized EEP-ROM cells, the .DB directive is preceded by the label text1. At the end of these EEPROM preparations, the .CSEG directive switches back to the code

```
;*********************************************************************
;* File Name          :ee_table.asm
;* Title              :Reading a table stored in EEPROM and display
;* Date               :09/06/97
;* Version            :1.0
;* Target MCU         :AT90S1200
;*
;* DESCRIPTION
;* Strings stored as a table in EEPROM are sent via RS232 to a display.
;*********************************************************************


;***** Directives
.nolist
.include "1200def.inc"
.list

.ESEG
text1:      ;Welcome to
            .db 87,101,108, 99,111,109,101, 32,116,111, 32
            ;Atmel's new AVR
            .db 65,116,109,101,108, 39,115, 32, 110,101,119, 32, 65, 86, 82
            ;micros.
            .db 32, 109,105, 99,114,111,115, 46, 13, 10, 00
.CSEG


; Registers used for EEPROM access (may be < r16)
; --------------------------------------------------------------------
.def        EEwtmp         = r0    ;temporary storage of address
.def        EErtmp         = r0    ;temporary storage of address
.def        EEdrd          = r0    ;result data byte
.def        EEdrd_s        = r1    ;result data byte
```

Listing 5-10
Table in EEPROM (ee_table.asm).   (Continued)

```
; Registers used for EEPROM access (must be >= r16)
; ------------------------------------------------------------------
.def        EEdwr          = r16     ;data byte to write to EEPROM
.def        EEawr          = r17     ;address to write to
.def        EEard          = r16     ;address to read from
.def        EEdwr_s        = r16     ;data to write

; Registers used for SW uart (must be >= r16)
; ------------------------------------------------------------------
.def        d_bcntr        = r23     ;bit counter
.def        d_srdel        = r24     ;delay loop variable
.def        d_txrx         = r25     ;async serial data byte transmit/
                                     ;received

; Main Program Register variables
; ------------------------------------------------------------------
.def        dly      = r17
.def        delcnt   = r18           ; Loop Counter
.def        temp     = r19

;***** Interrupt vector table
            rjmp     RESET            ; Reset handle
            reti                      ; External Interrupt0 handle
            reti                      ; Overflow0 Interrupt handle
            reti                      ; Analog Comparator Interrupt handle

;***** Includes

.include "eeprom.inc"
.include "uart.inc"

;***** Subroutines

DELAY:      clr      delcnt          ; Init Loop Counter
loop1:      dec      delcnt
            nop
            brne     loop1
            dec      dly
            brne     loop1
            ret
```

Listing 5-10
Continued

```
;***** Main

RESET:    rcall   d_sersetup    ; Setup UART pins
          ser     temp
          out     DDRB, temp
          out     PORTB,temp    ; Setup PortB

start:    ldi     temp, text1-1
          out     EEAR,temp     ; Set address of string

loop:     rcall   EERead_seq    ; get EEPROM data
          out     PORTB, EEdrd_s ; write to PortB
          mov     d_txrx,EEdrd_s ; and to transmit register
          rcall   d_writeser    ; Send character

          ldi     dly, $FF      ; Initialize delay of about 65 ms
          rcall   DELAY         ; Number of cycles = 1026 * del + 4

          tst     EEdrd_s       ; last character? (0-terminated
                                ; string)
          breq    start

          rjmp    loop          ; Repeat endless
```

Listing 5-10
Continued

memory for placing the interrupt vector table in the first program locations, followed by the rest of the code.

The assembler generates a separate hex file for programming the EEPROM. Knowing the structure of an Intel hex file makes it possible to patch this file for changing its contents. Listing 5-11 shows the contents of the generated Intel hex file.

```
:1000000057656C636F6D6520746F2041746D656C0E
:100010002773206E657720415652206D6963726F99
:05002000732E0D0A0023
:00000001FF
```

Listing 5-11
Contents of file ee_table.eep.

Example Programs

A description of the structure of Intel hex files and a hex-to-ASCII conversion table can be found in the Appendix. After some decoding work you will find the text string "Welcome to Atmel's new AVR micros" followed by CR, LF, and the 0-terminator again.

The source lines before the interrupt vector table define registers used by the include files and the main program itself. Only the used defines are necessary here, but all were defined simply by cut and paste. In the subroutine part, the delay subroutine used earlier was placed to slow down the output bit stream.

In the main part of this program, some initializations are done before an address is loaded in the address register for EEPROM access. To read the EEPROM cells, the sequential read subroutine EERead_seq from include file eeprom.inc is used. Based on Atmel's application note AVR100.asm, this include file was generated. Subroutines for random and sequential read/write access are implemented. The subroutines for sequential access start with incrementing the address counter. That is because the address register was loaded with the address text1-1 and not address text1 unchanged.

The byte read from EEPROM is output as a bit pattern to PortB and then sent to RS-232 using the subroutine d_writeser from the include file duart.inc. This procedure repeats until the character "0" is read out. The text string is built as a 0-terminated string. So, this received "0" marks the end of this string.

After output of the complete text string, the start address of the text string is reinitialized and the whole process repeats automatically (Figure 5-15).

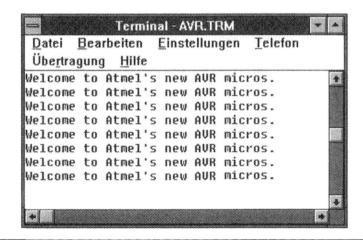

Figure 5-15
Output of text string.

AVR RISC Microcontroller Handbook

Security Enhancement in Data Transmission by CRC The RS-232-based data transmission in the two previous sections happened character by character without any provision for security or error detection. A correct data transmission between microcontrollers will be secured by a CRC (cyclic redundancy check). Because this method gives a high level of security, it is used in numerous communication protocols.

To calculate a CRC, the bits of the data byte are thought of as coefficients of a (possibly) very long polynomial. This polynomial will be divided by another polynomial; the result is the CRC. Clearly, the second polynomial is the key to security. The mathematical theory delivers some possibilities. For data in byte format, the following two polynomials are mostly used:

$$CRC\text{-}16: \quad x_{16} + x_{15} + x_2 + 1$$
$$CRC\text{-}CCITT: \quad x_{16} + x_{12} + x_2 + 1.$$

CRC-16 is mainly used in the United States, and CRC-CCITT is mainly used in Europe.

The program example in Listing 5-12 demonstrates the bytewise calculation of a CRC for a data package of the three bytes 12_H, 34_H, 56_H.

```
;********************************************************************
;* File Name           :crc.asm
;* Title               :CRC Check
;* Date                :09/20/97
;* Version             :1.0
;* Target MCU          :AT90S1200
;*
;* DESCRIPTION
;* Checking of data integrity by CRC
;********************************************************************

;***** Directives

.device at90s1200         ; Device is AT90S1200
.nolist
.include "1200def.inc"
.list

.equ    crcpoly_lo_byt  = $21    ; CRC-CCITT is $1021
.equ    crcpoly_hi_byte = $10

.def    temp_lo         = r0
.def    temp_hi         = r1
```

Listing 5-12
CRC calculation (crc.asm).   (Continued)

```
.def      temp            = r16
.def      byte            = r17
.def      crc_lo          = r18
.def      crc_hi          = r19
.def      crcpoly_lo      = r20
.def      crcpoly_hi      = r21
.def      count           = r22

;***** Interrupt vector table

          rjmp    RESET               ; Reset handle
          reti                        ; External Interrupt0 handle
          reti                        ; Overflow0 Interrupt handle
          reti                        ; Analog Comparator Interrupt handle

;***** Subroutines

calc_crc:
          eor     byte, crc_hi
          mov     crc_hi, byte
          ldi     count, 8
repeat:   mov     temp, crc_hi
          andi    temp, $80
          breq    shift_only
          lsl     crc_lo
          rol     crc_hi
          eor     crc_lo, crcpoly_lo
          eor     crc_hi, crcpoly_hi
          rjmp    next
shift_only:
          lsl     crc_lo
          rol     crc_hi
next:     dec     count
          brne    repeat
          ret

;***** Main

RESET:    ldi     crcpoly_lo, crcpoly_lo_byte ; Initialize CRC
                                             ; Polynom Lo Byte
          ldi     crcpoly_hi, crcpoly_hi_byte ; Initialize CRC
                                             ; Polynom Hi Byte

          clr     crc_lo          ; Clear CRC Lo Byte
          clr     crc_hi          ; Clear CRC Hi Byte

          ldi     byte, $12       ; Load 1. Byte
          rcall   calc_crc        ; Call CRC Calculation

          ldi     byte, $34       ; Load 2. Byte
          rcall   calc_crc        ; Call CRC Calculation
```

Listing 5-12
Continued

```
        ldi     byte, $56         ; Load 3. Byte
        rcall   calc_crc          ; Call CRC Calculation

        mov     temp_lo, crc_lo   ; Store Calculated CRC temporarily
        mov     temp_hi, crc_hi

        clr     crc_lo            ; Clear CRC Lo Byte
        clr     crc_hi            ; Clear CRC Hi Byte

        ldi     byte, $12         ; Load 1. Byte
        rcall   calc_crc          ; Call CRC Calculation

        ldi     byte, $34         ; Load 2. Byte
        rcall   calc_crc          ; Call CRC Calculation

        ldi     byte, $56         ; Load 3. Byte
        rcall   calc_crc          ; Call CRC Calculation

        mov     byte, temp_hi     ; Load CRC Hi Byte
        rcall   calc_crc          ; Call CRC Calculation

        mov     byte, temp_lo     ; Load CRC Lo Byte
        rcall   calc_crc          ; Call CRC Calculation

loop:   rjmp    loop              ; Repeat forever
```

Listing 5-12
Continued

In case of data transmission, the calculated 16-bit CRC will be attached to the (three) bytes sent already. The receiver gets enhanced data messages (two bytes more) and can calculate the CRC over the complete message again. If the result is not zero, the data transmission was corrupted.

The program in Listing 5-12 is prepared for the simulator. By ldi instructions, the three bytes are loaded and the CRC is calculated. For the order of bytes 12_H, 34_H, and 56_H, the calculated CRC is $DE61_H$.

In reality, the sending device would output the bytes 12_H, 34_H, 56_H, DE_H, and 61_H. The receiver had to calculate the CRC over these five bytes afterwards. In the program example just given, the calculated CRC is stored temporarily followed by a new CRC calculation over 5 bytes.

Simulating the CRC calculation in AVR Studio gives the possibility of setting a breakpoint for the endless loop. After the simulation of sending and receiving a series of bytes, the program will halt. In registers R1 and R0 the calculated CRCs of the sent bytes are stored, whereas in the registers R19 and R18, the recalculated CRCs of the received bytes is stored.

As can be seen from Figure 5-16, the calculated results indicate an error-free data transmission.

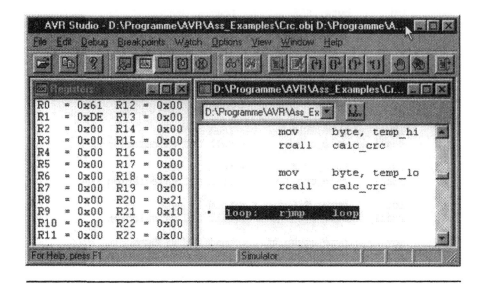

Figure 5-16
Simulated CRC calculation (`crc.asm`) in AVR Studio.

The subroutine `calc_crc` does all the work of CRC calculation. It is easy to include these features in applications where transmission of uncorrupted data is an important issue.

Software SPI The conditions for data transmission via SPI interface were explained in the section on the hardware of AVR microcontrollers. The AT90S1200 has no internal hardware to support serial communication with an SPI interface. To make SPI data exchange possible, an implementation of SPI in software is the only way.

In our software example, an EEPROM NM25C04 serves as part of an SPI data exchange. Before we look at the software, we make some remarks on the NM25C04 EEPROM. For further details, please look at the data sheet.

The NM25C04 is a 4096-bit SPI-compatible CMOS EEPROM designed for data storage in applications requiring both nonvolatile memory and in-system data updates. This EEPROM is well suited for applications using microcontrollers that support the SPI protocol for high-speed communication with peripheral devices via a serial bus to reduce pin count. Figure 5-17 shows the pinout of NM25C04.

The serial data transmission of this device requires four signal lines to control the device operation: Chip Select (CS), Clock (SCK), Serial Data In (SI), and Serial Data Out (SO). All programming cycles are completely self-timed and do not require an erase before WRITE. BLOCK WRITE protection is

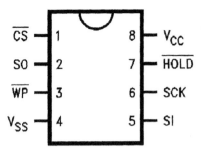

Figure 5-17
NM25C04 pinout.

provided by programming the STATUS REGISTER with one of four levels of write protection. Additionally, separate program enable and program disable instructions are provided for data protection.

Hardware data protection is provided by the /WP pin to protect against accidental data changes. The /HOLD pin allows the serial communication to be suspended without resetting the serial sequence. Neither feature is used in this example.

The features of the NM25C04 are, in summary:

- 4096 bits organized as 512 bytes
- Multiple chips on the same three-wire bus with separate chip select lines
- Self-timed programming cycle
- Simultaneous programming of 1 to 4 bytes at a time
- Status register can be polled during programming to monitor RDY/BUSY
- Write Protect (WP) pin and write disable instruction for both hardware and software write protection
- Block write protect feature to protect against accidental writes
- Endurance: 10^6 data changes
- Data retention greater than 40 years
- Packages available: 8-pin DIP or 8-pin SO

To understand the following software it is important to know how data is exchanged between the AVR microcontroller and the EEPROM used.

Data is accepted from NM25C04 only when /CS is active (/CS = Lo). The first byte transmitted after the chip is selected must contain the opcode that defines the operation to be performed. In the READ and WRITE instructions, the opcode also contains the address bit A8. The opcode for NM25C04 access is defined in the source text of the include file described next.

After an invalid code is received, no data is shifted into the NM25C04, and the SO data output pin remains at high impedance until a new /CS falling edge reinitializes the serial communication.

The required routines for SPI support are implemented as an include file suitable for all AVR microcontrollers. The primary subroutines call one further subroutine only, so the AT90S1200 with its hardware stack can work properly.

Listing 5-13 shows the source text of the include file sw_spi.inc built according to the same philosophy as the SPI driver for AVR microcontrollers with internal SPI support, explained later.

```
;*********************************************************************
;* File Name           :sw_spi.inc
;* Title               :Software SPI Subroutines
;* Author              :C.Kuehnel
;* Date                :10/25/97
;* Version             :1.0
;* Target MCU          :AT90S1200
;*
;* DESCRIPTION
;* Software SPI subroutines w/o interrupt.
;* Compatible with all AVR processors.
;*********************************************************************


; Primary routines:
; -----------------
; spi_setup:          Setup I/O bits for SPI. Must call prior to
;                     using the following routines.
; spi_write:          Write SPI device with addr in spi_addr and
;                     data byte to write in spi_data.
; spi_read:           Read SPI device with addr in spi_addr and data
;                     byte stored in spi_data.
; write_en:           Enables the EEPROM for writing. Enable before
;                     any write!
; write_di:           Disables the EEPROM for writing.
; block_protection:   Block Write Protection in EEPROM.
; wait_for_spi_ready: Wait for /RDY after write operation.
;
;
; 25C04 EEPROM OpCode Definitions
; -------------------------------------------------------------------
```

Listing 5-13
SPI Subroutines (sw_spi.inc).

```
.equ    WREN      = 0b00000110  ; Set Write Enable Latch
.equ    WRDI      = 0b00000100  ; Reset Write Enable Latch
.equ    RDSR      = 0b00000101  ; Read Status Register
.equ    WRSR      = 0b00000001  ; Write Status Register
.equ    READ_Lo   = 0b00000011  ; Read Data from lower memory area (A8=0)
.equ    READ_Hi   = 0b00001011  ; Read Data from upper memory area (A8=1)
.equ    WRITE_Lo  = 0b00000010  ; Write Data to lower memory area (A8=0)
.equ    WRITE_Hi  = 0b00001010  ; Write Data to upper memory area (A8=1)
.equ    NO_BL_PROT = 0b00000000 ; No Block Write Protection
.equ    BL2_PROT  = 0b00001000  ; Block Write Protection ($100..$1FF)
.equ    dummy     = $FF         ; Dummy byte
.equ    SCK       = 7
.equ    MISO      = 6
.equ    MOSI      = 5
.equ    SS        = 4
.equ    MSB       = 7


; Registers used by SPI subroutines,
; must be defined in calling program.
; --------------------------------------------------------------------
;    spi_opcode                 ; SPI OpCode
;    spi_addr                   ; SPI Memory Location
;    spi_data                   ; SPI Data
;    spi_buffer                 ; SPI Shift Register
;    bit_count                  ; Bit Counter

SPI_transmission:
        ldi       bit_count, 8
tra_bit:clc
        cbi       PORTB, MOSI
        sbrc      spi_buffer, MSB
        sbi       PORTB, MOSI
        sbi       PORTB, SCK
        lsl       spi_buffer
        sbic      PINB, MISO
        ori       spi_buffer, 1
        cbi       PORTB, SCK
        dec       bit_count
        brne      tra_bit
        ret
```

Listing 5-13
Continued

(*Continued*)

```
spi_setup:
    sbi         DDRB, SCK           ; SCK output
    cbi         PORTB, SCK          ; SCK Lo
    cbi         DDRB, MISO          ; MISO input
    sbi         DDRB, MOSI          ; MOSI output
    sbi         DDRB, SS        ; /SS output controls /CS of SPI peripheral
    sbi         PORTB, SS           ; /SS Hi
    ret

write_en:                       ; Set Write Enable Latch
    cbi         PORTB, SS           ; /CS Lo
    ldi         spi_buffer, WREN
    rcall       SPI_transmission    ; Write Enable
    sbi         PORTB, SS           ; /CS Hi
    ret

write_di:                       ; Reset Write Enable Latch
    cbi         PORTB, SS           ; /CS Lo
    ldi         spi_buffer, WRDI
    rcall       SPI_transmission    ; Write Disable
    sbi         PORTB, SS           ; /CS Hi
    ret

block_protection:               ; Set Block Write Protection
    cbi         PORTB, SS           ; /CS Lo
    ldi         spi_buffer, WRSR
    rcall       SPI_transmission    ; Write WRSR
    mov         spi_buffer, spi_data
    rcall       SPI_transmission    ; Write Status Register
    sbi         PORTB, SS           ; /CS Hi
    ret

wait_for_spi_ready:                     ; Wait until /RDY flag is cleared
    cbi         PORTB, SS
    ldi         spi_buffer, RDSR
    rcall       SPI_transmission    ; Read Status Register
    ldi         spi_buffer, dummy
rd1: rcall      SPI_transmission
    sbrc        spi_buffer, 0       ; if Bit0 (/RDY) is not cleared,
                                    ; read again
```

Listing 5-13
Continued

```
        rjmp       rd1
        sbi        PORTB, SS
        ret


spi_write:                              ; Write Data Byte to SPI address
        cbi        PORTB, SS
        ldi        spi_buffer, WRITE_Hi
        rcall      SPI_transmission
        mov        spi_buffer, spi_addr
        rcall      SPI_transmission
        mov        spi_buffer, spi_data
        rcall      SPI_transmission
        sbi        PORTB, SS
        ret


spi_read:                        ; Read Data byte from SPI address
        cbi        PORTB, SS
        ldi        spi_buffer, READ_Hi
        rcall      SPI_transmission
        mov        spi_buffer, spi_addr
        rcall      SPI_transmission
        ldi        spi_buffer, dummy
        rcall      SPI_transmission
        mov        spi_data, spi_buffer
        sbi        PORTB, SS
        ret
```

Listing 5-13
Continued

Because of the lack of hardware support for SPI in some AVR microcontrollers, the defined SPI buffer (spi_buffer) has special importance. All
data sent to the SPI device or received from this device is transferred through
this buffer with the subroutine SPI_transmission.

Any byte to be sent must be stored in the SPI buffer at first (Figure 5-18).
In a first step, the MSB will be sent to the SPI device (1) followed by a shift
operation (2) of 1 bit to the left. The last step is setting the LSB (3), if a Hi
level was received from the SPI device. This operation is repeated eight
times, until the byte placed into the SPI buffer at first is replaced by the received byte. Figure 5-18 shows this operation for 1-bit data exchange. All
subroutines accessing the NM25C04 use the subroutine SPI_transmission

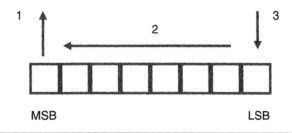

1
2
3

MSB
LSB

Figure 5-18
Bit operations in SPI buffer.

for data exchange. The registers spi_opcode, spi_addr, and spi_data were defined for compatibility between both SPI programs only.

We now understand as primary routines those implementing the requirements for a successful access to the NM25C04 device.

The data direction and default level of the involved I/O pins are defined by the subroutine spi_setup. The subroutine spi_read reads the contents of the addressed memory cell, whereas the subroutine spi_write programs an addressed memory cell. Because of the use of commands READ_Hi and WRITE_Hi, both subroutines address a memory location in the the upper memory area. For access to the lower area, use the commands READ_Lo and WRITE_Lo in the corresponding subroutines.

The duration of one programming cycle reaches up to 5 ms and is signaled by the /BSY flag. The subroutine wait_for_spi_ready queries the status register of NM25C04 and waits until the end of programming. For reading a data byte, a dummy byte must always be sent.

To test the include file sw_spi.inc, the program example spi4all.asm calls the subroutines described. Listing 5-14 shows the source text of program example spi4all.asm. This program example was written for the AT90S1200, but the test was done with an AT90S8515 without using its internal SPI hardware. The reason for this procedure was the additional use of an RS-232 for debugging with an external monitor. For working with an AT90S1200, some changes in the source must be made. All such lines are marked in the source text.

The idea of this program is to write a byte to any location in EEPROM and to read back. Register period is set depending on the comparison result of both bytes. If you get a quickly flickering LED at PB0, an error has occurred. If the LED flickers slowly, the read byte was equal to the byte formerly written.

```
;*********************************************************************
;* File Name       :spi4all.asm
;* Title           :SPI Test
;* Author          :C.Kuehnel
;* Date            :10/25/97
;* Version         :1.0
;* Target MCU      :AT90S1200 (tested with AT90S8515 w/o internal SPI)
;*
;* DESCRIPTION
;* SPI Test w/o interrupt.
;* Compatible with all AVR processors—no SPI hardware required.
;*********************************************************************

;***** Directives
.device    at90s8515          ; !!! must be changed for AT90S1200
.nolist
.include "8515def.inc"        ; !!! must be changed for AT90S1200
.list

; Primary routines:
; ------------------
; spi_setup:          Setup I/O bits for SPI. Must call prior to
;                     using the following routines.
; spi_write:          Write SPI device with addr in spi_addr and
;                     data byte to write in spi_data.
; spi_read:           Read SPI device with addr in spi_addr and data
;                     byte stored in spi_data.
; write_en:           Enables the EEPROM for writing. Enable before
;                     any write!
; write_di:           Disables the EEPROM for writing.
; block_protection:   Block Write Protection in EEPROM.
; wait_for_spi_ready: Wait for /RDY after write operation.
;

.equ    EE_addr    = $CC     ; Memory location
.equ    EE_data    = $44     ; Data to save

; Main Program Register variables
; ----------------------------------------------------------------
.def    temp       = r16
.def    period     = r17     ; Blink frequency
```

Listing 5-14
Reading and writing SPI EEPROM NM25C04 (spi4all.asm).

(*Continued*)

```
.def    dly        = r18
.def    delcnt     = r19    ; Loop Counter
.def    spi_opcode = r20    ; SPI OpCode
.def    spi_addr   = r21    ; SPI Memory Location
.def    spi_data   = r22    ; SPI Data
.def    spi_buffer = r23    ; SPI Shift Register
.def    bit_count  = r24    ; Bit Counter

;***** Interrupt vector table ; !!! must be changed for any other AVR
        rjmp        RESET    ; Reset handle
        reti                 ; External Interrupt0 handle
        reti                 ; External Interrupt1 handle
        reti                 ; T/C1 Capture Event Interrupt handle
        reti                 ; T/C1 CompareA Interrupt handle
        reti                 ; T/C1 CompareB Interrupt handle
        reti                 ; T/C1 Overflow Interrupt handle
        reti                 ; T/C0 Overflow Interrupt handle
        reti                 ; SPI Transfer Complete Interrupt handle
        reti                 ; UART Rx Complete Interrupt handle
        reti                 ; UART Data Register Empty Interrupt handle
        reti                 ; UART Tx Complete Interrupt handle
        reti                 ; Analog Comparator Interrupt handle

;***** Includes

.include "sw_spi.inc"
.include "hw_uart.inc"       ; !!! only for debugging using UART

;***** Subroutines

DELAY: clr       delcnt               ; Init Loop Counter
loop1: dec       delcnt
       nop
       brne      loop1
       dec       dly
       brne      loop1
       ret

;***** Main

RESET: ldi       temp, LOW(RAMEND)    ; !!! must be changed for
                                      ; AT90S1200
```

Listing 5-14
Continued

```
out         SPL, temp              ; Initialize SPL
ldi         temp, HIGH(RAMEND)
out         SPH, temp              ; Initialize SPH

sbi         DDRB, PB0              ; PB0 output
rcall       uart_init              ; Initialize UART, cancel for
                                   ; AT90S1200
rcall       spi_setup

ldi         period, $FF            ; Initalize period
                                   ; (FF ok | 60 for error)

;ldi        spi_data, BL2_PROT     ; Block Write Protection
                                   ; from $100 to $1FF
ldi         spi_data, NO_BL_PROT   ; No Block Write Protection
rcall       write_en
rcall       block_protection       ; Setup Status Register Bits
                                   ; in 25C04
rcall       wait_for_spi_ready

ldi         spi_addr, EE_addr
ldi         spi_data, EE_data
rcall       write_en
rcall       spi_write              ; Store Byte in 25C04
rcall       wait_for_spi_ready

clr         spi_data

ldi         spi_addr, EE_addr
rcall       spi_read               ; Read back stored byte

mov         temp, spi_data         ; Send it to terminal for test
rcall       putc                   ; !!! only for debugging
                                   ; using UART

ldi         temp, EE_data
cpse        spi_data, temp         ; Compare read back byte
                                   ; with byte written
ldi         period, $60
```

Listing 5-14
Continued

(*Continued*)

```
loop:   sbi         PORTB, PB0
        mov         dly, period             ; Initialize delay controlled
                                            ; by period
        rcall       DELAY
        cbi         PORTB, PB0
        mov         dly, period             ; Initialize delay controlled
                                            ; by period
        rcall       DELAY
        rjmp        loop                    ; Repeat forever
```

Listing 5-14
Continued

The write and read operations point to the upper memory area (A8=1). You can test the block protection by recommenting the instruction ;ldi spi_data, BL2_PROT, and commenting the instruction ldi spi_data, NO_BL_PROT afterwards. In this case, any writing in this memory area is impossible.

If you changed the EE_data before the trial of writing in a protected area, the comparision must always fail.

5.1.2 Assembler Programs for the AT90S8515

The Assembler programs for the AT90S2313, AT90S4414, and the AT90S8515 differ from programs for AT90S1200 not only because of enhanced peripherals. These peripherals can offload the CPU. Furthermore, the AT90S1200 works on a three-level hardware stack that must not be initialized.

First Test with the AT90S8515   Even for a microcontroller with more internal resources, it is best to a start with a simple test first. Let us modify the program mc_test1.asm to the AT90S8515 requirements. Listing 5-15 shows the slightly changed microcontroller test program adapted to AT90S8515.

We find the following changes with reference to the microcontroller test program for the AT90S1200:

- Changed device instruction (.device at90s8515)
- Changed include file (.include "8515def.inc")
- Enhanced interrupt vector table
- Required software stack initialization after reset

```
;**********************************************************************
;* File Name            :mc_test2.asm
;* Title                :Microcontroller Test
;* Date                 :09/28/97
;* Version              :1.0
;* Target MCU           :AT90S8515
;*
;* DESCRIPTION
;* Test of Timer0 Overflow and External Interrupt on AVR Evaluation Board
;* Oscillator frequency CK is 4 MHz (T = 0.25 us)
;**********************************************************************


;***** Directives

.device  at90s8515       ; Device is AT90S8515
.nolist
.include "8515def.inc"
.list
.listmac

.def     temp    = r16
.def     byte    = r17
.def     reload  = r18


;***** Macros

.MACRO   SBIs_HR                 ; Set Bits in I/O Register 32 up
         in      temp,@0
         ori     temp, @1
         out     @0, temp
.ENDMACRO


;***** Interrupt vector table

         rjmp    RESET           ; Reset handle
         rjmp    EX_INT0         ; External Interrupt0 handle
         reti                    ; External Interrupt1 handle
         reti                    ; T/C1 Capture Event Interrupt handle
         reti                    ; T/C1 CompareA Interrupt handle
         reti                    ; T/C1 CompareB Interrupt handle
         reti                    ; T/C1 Overflow Interrupt handle
         rjmp    OVF0            ; T/C0 Overflow Interrupt handle
         reti                    ; SPI Transfer Complete Interrupt handle
```

Listing 5-15
Test program (mc_test2.asm).   (*Continued*)

```
        reti                    ; UART Rx Complete Interrupt handle
        reti                    ; UART Data Register Empty Interrupt
                                ; handle
        reti                    ; UART Tx Complete Interrupt handle
        reti                    ; Analog Comparator Interrupt handle

;***** Interrupt handlers

EX_INT0: swap   reload          ; Rotate Reload
        reti

OVF0:   out     PORTB, byte     ; Output Bit Pattern
        rol     byte            ; Rotate Bit Pattern
        out     TCNT0, reload   ; Reload Timer/Counter0
        reti

;***** Main

RESET:  ldi     temp, LOW(RAMEND)
        out     SPL, temp               ; Initialize SPL
        ldi     temp, HIGH(RAMEND)
        out     SPH, temp               ; Initialize SPH

        ldi     reload,$07      ; Initialize Reload Value

        sec
        ldi     byte, $FE
        out     PORTB,byte      ; Initialize PortB

        ser     temp
        out     DDRB,temp       ; PORTB = all outputs

        ldi     temp, 0b00000101
        out     TCCR0, temp     ; Prescaler CK/1024 => 256 us

        ldi     temp, TOV0<<1
        out     TIMSK, temp     ; T/C0 Interrupt Enable

        SBIs_HR MCUCR,0b00000010; Interrupt on falling edge of INT0

        SBIs_HR GIMSK,0b01000000; External Interrupt0 Enable

        sei                     ; Global Interrupt Enable

loop:   rjmp    loop            ; Repeat forever
```

Listing 5-15
Continued

The remaining functions are the same as those described in the section on the AT90S1200 microcontroller. To test the external interrupt capability, Pin2 of PortD again serves as input for the external interrupt INT0. The circuitry for this program example is the same as that shown in Figure 5-1. The generated bit pattern will be output on PortB periodically, and a pressed key must change the refreshing period for this output.

Pulse-Width Modulation   Using the enhanced capabilities of Timer/Counter1 relieves the CPU from PWM tasks. Furthermore, the resolution is enhanced to 10-bit maximum.

The pulse signals shown in Figure 5-19 can be generated by the Timer/Counter1 in PWM mode. In 10-bit mode the Timer/Counter1 makes 1024 different duty cycles possible. The resulting wave form available on pin OCR1A and/or OCR1B can be integrated by a resistor and a capacitor to get an analogous voltage.

To show the results, we first look at Table 5-3 and Figure 5-20. Both figures list the voltage generated by pulse-width modulation and measured with a digital voltmeter. One single step is nominal 4.75 mV. The overall linearity again looks very good.

The program for pulse-width generation with Timer/Counter1 is quite simple. Listing 5-16 shows this program example. The main task is a suitable initialization of Timer/Counter1.

In the subroutine `pwm1_init`, Timer/Counter1 is set in 10-bit PWM mode with a PWM frequency of 500 kHz. The actual duty cycle (000 .. 3FF$_H$) is initialized outside this subroutine. The duty cycle in program `pwm1.asm` is fixed.

To get all values listed in Table 5-3, Hi and Lo bytes of register OCR1X were changed in the source text. After newly assembling the source and programming the AT90S8515, the new PWM output voltage could be measured on the relating I/O pins. This procedure sounds more troublesome than it is, at least in a Windows environment with Assembler and Programmer opened permanently.

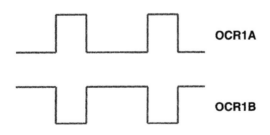

Figure 5-19
PWM wave form.

Table 5-3
Measured voltage of PWM with Timer/Counter1.

Byte	$000	$010	$020	$030	$040	$050	$060	$070	$080	$090	$0A0	$0B0	$0C0	$0D0	$0E0	$0F0
OC1A	.001	.078	.156	.234	.311	.389	.467	.544	.622	.700	.777	.855	.932	1.01	1.08	1.15
OC1B	4.93	4.86	4.78	4.70	4.62	4.55	4.47	4.39	4.32	4.24	4.16	4.08	4.01	3.93	3.85	3.77
Byte	$100	$110	$120	$130	$140	$150	$160	$170	$180	$190	$1A0	$1B0	$1C0	$1D0	$1E0	$1F0
OC1A	1.23	1.31	1.39	1.46	1.54	1.62	1.69	1.77	1.85	1.93	2.00	2.08	2.16	2.24	2.31	2.39
OC1B	3.70	3.62	3.54	3.47	3.39	3.31	3.23	3.16	3.08	3.00	2.93	2.85	2.77	2.69	2.61	2.54
Byte	$200	$210	$220	$230	$240	$250	$260	$270	$280	$290	$2A0	$2B0	$2C0	$2D0	$2E0	$2F0
OC1A	2.47	2.54	2.62	2.70	2.78	2.85	2.93	3.01	3.08	3.16	3.24	3.32	3.39	3.47	3.55	3.63
OC1B	2.46	2.39	2.31	2.23	2.15	2.08	2.00	1.92	1.85	1.77	1.69	1.61	1.54	1.46	1.38	1.30
Byte	$300	$310	$320	$330	$340	$350	$360	$370	$380	$390	$3A0	$3B0	$3C0	$3D0	$3E0	$3F0
OC1A	3.70	3.78	3.86	3.93	4.01	4.09	4.17	4.24	4.32	4.40	4.47	4.55	4.63	4.71	4.78	4.86
OC1B	1.23	1.15	1.07	1.007	.929	.851	.774	.696	.618	.541	.463	.385	.308	.230	.152	.075

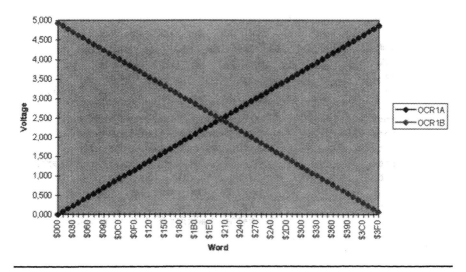

Figure 5-20
Voltage output on OCR1A and OCR1B by pulse-width modulation.

```
;*********************************************************************
;* File Name          :pwm1.asm
;* Title              :Pulse Width Modulation

;* Author             :C.Kuehnel
;* Date               :10/15/97
;* Version            :1.0
;* Target MCU         :AT90S8515
;*
;* DESCRIPTION
;* Pulse width modulation by Timer/Counter1
;*********************************************************************

;***** Directives
.device    at90s8515
.nolist
.include "8515def.inc"
.list

; Main Program Register variables
; -------------------------------------------------------------------
.def       temp            = r16
```

Listing 5-16
Pulse-width generation by Timer/Counter1 (pwm1.asm).

(*Continued*)

```
;***** Interrupt vector table

        rjmp    RESET           ; Reset handle
        reti                    ; External Interrupt0 handle
        reti                    ; External Interrupt1 handle
        reti                    ; T/C1 Capture Event Interrupt handle
        reti                    ; T/C1 CompareA Interrupt handle
        reti                    ; T/C1 CompareB Interrupt handle
        reti                    ; T/C1 Overflow Interrupt handle
        reti                    ; T/C0 Overflow Interrupt handle
        reti                    ; SPI Transfer Complete Interrupt handle
        reti                    ; UART Rx Complete Interrupt handle
        reti                    ; UART Data Register Empty Interrupt handle
        reti                    ; UART Tx Complete Interrupt handle
        reti                    ; Analog Comparator Interrupt handle

;***** Subroutines

pwm1_init:                      ; Initializes T/C1 as 10-bit PWM
        ldi     temp, 0b10110011 ; Timer/Counter1 is a 10-bit PWM
        out     TCCR1A, temp
        ldi     temp, 0b00000010 ; CK/8 => PWM frequency is 500 kHz
        out     TCCR1B, temp
        ret

pwm1_disable:
        clr     temp            ; Disable T/C1
        out     TCCR1A, temp
        ret

;***** Main

RESET:  ldi     temp, LOW(RAMEND)
        out     SPL, temp       ; Initialize SPL
        ldi     temp, HIGH(RAMEND)
        out     SPH, temp       ; Initialize SPH
        sbi     DDRD, PD5       ; PD5 is output

        ldi     temp, $03       ; Load the Hi Byte of OCR1X (00..3F)
        out     OCR1AH, temp
        out     OCR1BH, temp

        ldi     temp, $F0       ; Load the Lo Byte of OCR1X (00..FF)
        out     OCR1AL, temp
        out     OCR1BL, temp

        rcall   pwm1_init

loop:   rjmp    loop            ; Repeat forever
```

Listing 5-16
Continued

Serial Communication with the Internal UART Serial communication
—a very important function in all microcontroller applications—is very easy
with the use of an internal UART.

In practical designs we have to distinguish among different implementations of serial communication:

(1) Interrupt-driven data exchange for receive and transmit
(2) Interrupt-driven data exchange for receive only
(3) Data exchange without UART interrupts

All three kinds of implementation can be found in real applications and have
their special advantages and disadvantages.

An interrupt-driven data exchange for receive and transmit is a sophisticated solution for handling high bidirectional data traffic. The data received
and to be transmitted must be temporarily saved in buffers. The interrupt handler has to manage these buffers by read and write pointer adjustments, and so
forth, on one side. The application program has to manage these buffers from
the other side.

Because data is received and transmitted in the background, no CPU resources are used up for the communication tasks. Going without interrupts,
the CPU has to handle the communication. With an internal UART, this means
reading and writing the UART I/O data registers and often waiting until the
UART status register signalizes readiness for data exchange. In the following
program example, based on this simple implementation, we will see that only
half-duplex data exchange is possible.

Because of its easier implementation, the following program example works
without any UART interrupts. Listing 5-17 shows a simple test program for
RS-232-based serial communication.

```
;*******************************************************************
;* File Name          :uart.asm
;* Title              :UART based RS232 communication
;* Date               :09/28/97
;* Version            :1.0
;* Target MCU         :AT90S8515
;*
;* DESCRIPTION
;* Test the UART based RS232 communication
;*******************************************************************
```

Listing 5-17
UART test program (uart.asm).   (*Continued*)

```
;***** Directives
.device    at90s8515
.nolist
.include "8515def.inc"
.list

.def    temp = r16

;***** Interrupt vector table

        rjmp    RESET           ; Reset handle
        reti                    ; External Interrupt0 handle
        reti                    ; External Interrupt1 handle
        reti                    ; T/C1 Capture Event Interrupt handle
        reti                    ; T/C1 CompareA Interrupt handle
        reti                    ; T/C1 CompareB Interrupt handle
        reti                    ; T/C1 Overflow Interrupt handle
        reti                    ; T/C0 Overflow Interrupt handle
        reti                    ; SPI Transfer Complete Interrupt handle
        reti                    ; UART Rx Complete Interrupt handle
        reti                    ; UART Data Register Empty Interrupt handle
        reti                    ; UART Tx Complete Interrupt handle
        reti                    ; Analog Comparator Interrupt handle

;***** Subroutines

uart_init:
        ldi     temp, 25    ; Baudrate 9600 @ 4MHz Clock
;       ldi     temp, 12    ; Baudrate 19200 @ 4MHz Clock
        out     UBRR, temp
        ret

getc:   sbi     UCR,  RXEN  ; Rx Enable
        sbis    USR,  RXC   ; Wait for character
        rjmp    getc
        in      temp, UDR   ; Read character
        cbi     UCR,  RXEN  ; Rx Disable
        ret

putc:   sbi     UCR,  TXEN  ; Tx Enable
        sbis    USR,  UDRE  ; Wait until Data Register is empty
        rjmp    putc
        out     UDR,  temp  ; Write character
        cbi     UCR,  TXEN  ; Tx Disable
        ret
```

Listing 5-17
Continued

```
;***** Main

RESET: ldi     temp, LOW(RAMEND)
       out     SPL,  temp    ; Initialize SPL
       ldi     temp, HIGH(RAMEND)
       out     SPH,  temp    ; Initialize SPH
       rcall   uart_init     ; Initialize UART

loop:  rcall   getc          ; Read character from RS232
       inc     temp
       rcall   putc          ; Echo back incremented character
       rjmp    loop          ; Repeat endless
```

Listing 5-17
Continued

The UART test program contains three important subroutines:

• uart_init is responsible for baud rate setup
• getc waits until one character is received and reads it from register UDR
• putc writes one character to register UDR for sending

The main part of the program is an endless loop waiting for one character received via RS-232, incrementing this character as an example of data handling, and sending the incremented character via RS-232 back to the sender.

For testing the program uart.asm the COM port of a PC was connected to the second RS-232 (DB9) connector on Atmel's evaluation board. From Window's terminal program the character sequence "0," "1," "2," up to "9" was sent to the microcontroller running uart.asm. The terminal was waiting for the echoed sequence "1," "2," "3," ... to ":." Figure 5-21 shows the logged data traffic.

Because of the enabled echo, we first see the sent character, followed by the character that was sent back. So, two characters always build a character pair (sent character and sent-back character), as shown in Figure 5-21.

For further use of the defined subroutines, an include file hw_uart.inc was built from program uart.asm.

Tables in Code Area   All microcontrollers of the AVR family, with the exception of the AT90S12xx family, have the lpm instruction for reading from code area. In many microcontroller applications, the user dialog is stored as ASCII strings in code area.

Figure 5-21
Data traffic between terminal and program `uart.asm`.

In the next program example, a string is stored in code area. The program accesses this string character-by-character and sends these characters via RS-232 to a terminal for display. Listing 5-18 shows table reading in code area and data transmission of read characters.

```
;******************************************************************
;* File Name        :cc_table.asm
;* Title            :Reading a table stored in code area and display
;* Date             :09/28/97
;* Version          :1.0
;* Target MCU       :AT90S8515
;*
;* DESCRIPTION
;* Strings stored as a table in program memory
;* are sent via RS232 to a display.
;******************************************************************

;***** Directives
.device    at90s8515
.nolist
.include "8515def.inc"
.list
```

Listing 5-18
Reading a table strored in code area (`cc_table.asm`).

```
; Main Program Register variables
; -------------------------------------------------------------------
.def    temp    = r16
.def    dly     = r17
.def    delcnt  = r18             ; Loop Counter


;***** Interrupt vector table

        rjmp    RESET           ; Reset handle
        reti                    ; External Interrupt0 handle
        reti                    ; External Interrupt1 handle
        reti                    ; T/C1 Capture Event Interrupt handle
        reti                    ; T/C1 CompareA Interrupt handle
        reti                    ; T/C1 CompareB Interrupt handle
        reti                    ; T/C1 Overflow Interrupt handle
        reti                    ; T/C0 Overflow Interrupt handle
        reti                    ; SPI Transfer Complete Interrupt handle
        reti                    ; UART Rx Complete Interrupt handle
        reti                    ; UART Data Register Empty Interrupt handle
        reti                    ; UART Tx Complete Interrupt handle
        reti                    ; Analog Comparator Interrupt handle

;***** Table in Code Area (behind interrupt vector table)

text1:  ;Welcome to A
        .db     87,101,108, 99,111,109,101, 32,116,111, 32, 65
        ;tmel's new AVR
        .db     116,109,101,108, 39,115, 32,110,101,119, 32, 65, 86, 82
        ;micros.
        .db     32,109,105, 99,114,111,115, 46, 13, 10, 00

;***** Includes

.include "hw_uart.inc"

;***** Subroutines

DELAY:  clr     delcnt          ; Init Loop Counter
loop1:  dec     delcnt
        nop
        brne    loop1
        dec     dly
        brne    loop1
        ret
```

Listing 5-18
Continued

(Continued)

```
;***** Main

RESET:      ldi     temp, LOW(RAMEND)
            out     SPL, temp       ; Initialize SPL
            ldi     temp, HIGH(RAMEND)
            out     SPH, temp       ; Initialize SPH

            rcall   uart_init       ; Initialize UART
            ser     temp
            out     DDRB, temp
            out     PORTB,temp      ; Setup PortB

start:      ldi     ZL, LOW(text1<<1)
            ldi     ZH, HIGH(text1<<1); Load Z with address of string

loop:       lpm                     ; get data from program memory
            inc     ZL
            out     PORTB, R0       ; write to PortB
            mov     temp, R0        ; and to transmit register
            rcall   putc            ; Send character

            ldi     dly, $40        ; Initialize delay of about 16 ms
            rcall   DELAY           ; Number of cycles = 1026 * del + 4

            tst     R0              ; last character? (0-terminated string)
            breq    start
            rjmp loop               ; Repeat endless
```

Listing 5-18
Continued

Starting at the label text1, the string "Welcome to Atmel's new AVR micros." followed by CR, LF, and a zero byte are stored. To find the end-of-string, we use a zero for terminating, as is usual in C programs (0-terminated string).

When we use the directive .DB for placing characters in code area, a peculiarity must be taken into consideration. If the expressionlist contains more than one expression, the expressions are packed so that 2 bytes are placed in each program-memory word. If the expressionlist contains an odd number of expressions, the last expression will be placed in a program-memory word of its own, even if the next line in the assembly code contains a .DB directive. Because of this pecularity of the directive .DB, only the last line has an odd number of characters.

AVR RISC Microcontroller Handbook

How the string is stored in code memory can be examined with AVR Studio. Figure 5-22 shows the contents of the first 32 words in code memory. The text string starts at location $0D_H$ end ends at location $1F_H$ with the zero byte. The order of bytes sometimes leads to confusion.

In Listing 5-18, you can see that the access to data bytes or characters in code memory with the instruction lpm, together with addressing by Z register, takes place without any problems.

Serial Data Exchange with SPI   Section 5.1.1 described a software SPI for AVR microcontrollers without internal SPI hardware. This section will show the changes for AVR microcontrollers supporting the SPI directly.

As with the software SPI, the corresponding subroutines are implemented with the same names in the include file hw_spi.inc. Listing 5-19 shows the source text of this include file.

You can compare both include files directly. Only the subroutine SPI_transmission has no equivalent. The instructions out SPDR, *register* and in *register*, SPDR handle the data exchange.

Because the data exchange takes place in the background, one must wait with the next data to send until all bits are transmitted over the SPI lines. The end of data transmission is signaled by the SPI Interrupt flag, SPIF. Querying this flag and waiting until end of data transmission is indicated and avoids overwriting data.

The main program for testing the EEPROM access spi.asm is shown in Listing 5-20. It is quite similar to the program spi4all.asm and has no new peculiarities.

Software I²C   The I²C bus is a two-wire communication interface and allows synchronous bidirectional data transmission between transmitter and receiver using SCL (clock) and SDA (data I/O) lines. Peripheral devices with an I²C bus interface can, thus, enhance the I/O capabilities of microcontrollers with limited I/O resources.

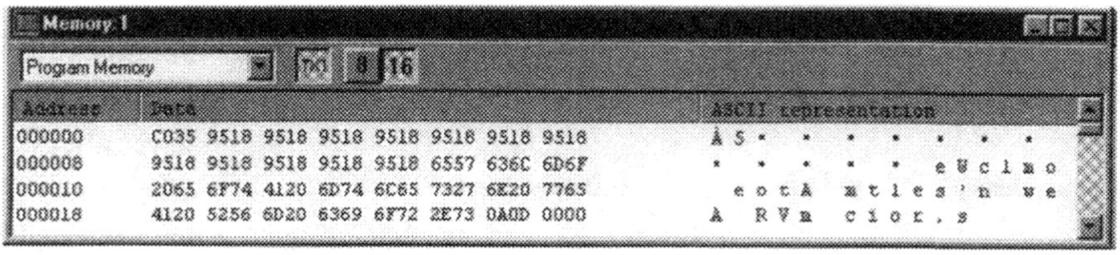

Figure 5-22
Table in code area (screenshot from AVR Studio).

```
;****************************************************************************
;* File Name          :hw_spi.inc
;* Title              :SPI Routinen
;* Author             :C.Kuehnel
;* Date               :10/25/97
;* Version            :1.0
;* Target MCU         :AT90S8515
;*
;* DESCRIPTION
;* SPI subroutines w/o interrupt.
;* Compatible with AVR processors including SPI hardware.
;****************************************************************************

; Primary routines:
; -----------------
; spi_setup:          Setup I/O bits for SPI. Must call prior to
;                     using the following routines.
; spi_write:          Write SPI device with addr in spi_addr and
;                     data byte to write in spi_data.
; spi_read:           Read SPI device with addr in spi_addr and
;                     data byte stored in spi_data.
; write_en:           Enables the EEPROM for writing. Enable
;                     before any write!
; write_di:           Disables the EEPROM for writing.
; block_protection:   Block Write Protection in EEPROM.
; wait_for_spi_ready: Wait for /RDY after write operation.
;
; 25C04 EEPROM OpCode Definitions
; ----------------------------------------------------------------------
.equ  WREN      = 0b00000110   ; Set Write Enable Latch
.equ  WRDI      = 0b00000100   ; Reset Write Enable Latch
.equ  RDSR      = 0b00000101   ; Read Status Register
.equ  WRSR      = 0b00000001   ; Write Status Register
.equ  READ_Lo   = 0b00000011   ; Read Data from lower memory area (A8=0)
.equ  READ_Hi   = 0b00001011   ; Read Data from upper memory area (A8=1)
.equ  WRITE_Lo  = 0b00000010   ; Write Data to lower memory area (A8=0)
.equ  WRITE_Hi  = 0b00001010   ; Write Data to upper memory area (A8=1)
.equ  NO_BL_PROT = 0b00000000  ; No Block Write Protection
.equ  BL2_PROT  = 0b00001000   ; Block Write Protection ($100..$1FF)

.equ  dummy     = $FF          ; Dummy byte
```

Listing 5-19
SPI subroutines (spi.inc).

```
; Registers used by SPI subroutines,
; must be defined in calling program.
; -----------------------------------------------------------------------
;           spi_opcode                  ; SPI OpCode
;           spi_addr                    ; SPI Memory Location
;           spi_data                    ; SPI Data


spi_setup:
            sbi     DDRB, PB7           ; SCK output
            cbi     DDRB, PB6           ; MISO input
            sbi     DDRB, PB5           ; MOSI output
            sbi     DDRB, PB4    ; /SS output controls /CS of SPI peripheral
            sbi     PORTB, PB4          ; /SS Hi
            ldi     temp, 0b01010100    ; MSB First, CPOL=1, CPHA=0, SCK=CLK/4
            out     SPCR, temp          ; Initialize SPI Control register
            ret


write_en:                       ; Set Write Enable Latch
            cbi     PORTB, PB4          ; /CS Lo
            ldi     spi_opcode, WREN
            out     SPDR, spi_opcode    ; Write Enable
we1:        SBIS    SPSR, SPIF
            rjmp    we1
            sbi     PORTB, PB4          ; /CS Hi
            ret


write_di:                       ; Reset Write Enable Latch
            cbi     PORTB, PB4          ; /CS Lo
            ldi     spi_opcode, WRDI
            out     SPDR, spi_opcode    ; Write Disable
wd1:        SBIS    SPSR, SPIF
            rjmp    wd1
            sbi     PORTB, PB4          ; /CS Hi
            ret


block_protection:               ; Set Block Write Protection
            cbi     PORTB, PB4          ; /CS Lo
            ldi     spi_opcode, WRSR
            out     SPDR, spi_opcode    ; Write WRSR
bp1:        SBIS    SPSR, SPIF
            rjmp    bp1
            out     SPDR, spi_data      ; Write Status Register
```

Listing 5-19
Continued

(*Continued*)

```
bp2:     SBIS    SPSR, SPIF
         rjmp    bp2
         sbi     PORTB, PB4      ; /CS Hi
         ret


wait_for_spi_ready:              ; Wait until /RDY flag is cleared
         cbi     PORTB, PB4
         ldi     spi_opcode, RDSR
         out     SPDR, spi_opcode ; Read Status Register
rd1:     SBIS    SPSR, SPIF
         rjmp    rd1
         ldi     spi_data, dummy
rd3:     out     SPDR, spi_data
rd2:     SBIS    SPSR, SPIF
         rjmp    rd2
         sbic    SPDR, 0         ; if Bit0 (/RDY) is not cleared, read again
         rjmp    rd3
         sbi     PORTB, PB4
         ret


spi_write:                       ; Write Data Byte to SPI address
         cbi     PORTB, PB4
         ldi     spi_opcode, WRITE_Hi
         out     SPDR, spi_opcode
w1:      SBIS    SPSR, SPIF
         rjmp    w1
         out     SPDR, spi_addr
w2:      SBIS    SPSR, SPIF
         rjmp    w2
         out     SPDR, spi_data
w3:      SBIS    SPSR, SPIF
         rjmp    w3
         sbi     PORTB, PB4
         ret


spi_read:                        ; Read Data byte from SPI address
         cbi     PORTB, PB4
         ldi     spi_opcode, READ_Hi
         out     SPDR, spi_opcode
r1:      SBIS    SPSR, SPIF
         rjmp    r1
         out     SPDR, spi_addr
```

Listing 5-19
Continued

AVR RISC Microcontroller Handbook

```
r2:      SBIS    SPSR, SPIF
         rjmp    r2
         ldi     spi_data, dummy
         out     SPDR, spi_data
r3:      SBIS    SPSR, SPIF
         rjmp    r3
         in      spi_data, SPDR
         sbi     PORTB, PB4
         ret
```

Listing 5-19
Continued

```
;************************************************************************
;* File Name           :spi.asm
;* Title               :SPI Test
;* Author              :C.Kuehnel
;* Date                :10/25/97
;* Version             :1.0
;* Target MCU          :AT90S8515
;*
;* DESCRIPTION
;* SPI Test w/o interrupt.
;* Compatible with AVR processors including SPI hardware.
;************************************************************************


;***** Directives
.device     at90s8515
.nolist
.include "8515def.inc"
.list


; Primary routines:
; -----------------
; spi_setup: Call to setup I/O bits for SPI. Must call prior to using the
;            following routines.
; spi_write: Call to write SPI device with byte to send in spi_buffer.
; spi_read:  Call to read SPI device, byte returned in spi_buffer.
```

Listing 5-20
Reading and writing SPI EEPROM NM25C04 (spi.asm).

(*Continued*)

```
.equ    EE_addr = $AA           ; Memory location
.equ    EE_data = $44           ; Data to save

; Main Program Register variables
; ----------------------------------------------------------------------
.def    temp        = r16
.def    period      = r17   ; Blink frequency
.def    dly         = r18
.def    delcnt      = r19   ; Loop Counter
.def    spi_opcode  = r20   ; SPI OpCode
.def    spi_addr    = r21   ; SPI Memory Location
.def    spi_data    = r22   ; SPI Data

;***** Interrupt vector table

        rjmp    RESET           ; Reset handle
        reti                    ; External Interrupt0 handle
        reti                    ; External Interrupt1 handle
        reti                    ; T/C1 Capture Event Interrupt handle
        reti                    ; T/C1 CompareA Interrupt handle
        reti                    ; T/C1 CompareB Interrupt handle
        reti                    ; T/C1 Overflow Interrupt handle
        reti                    ; T/C0 Overflow Interrupt handle
        reti                    ; SPI Transfer Complete Interrupt handle
        reti                    ; UART Rx Complete Interrupt handle
        reti                    ; UART Data Register Empty Interrupt handle
        reti                    ; UART Tx Complete Interrupt handle
        reti                    ; Analog Comparator Interrupt handle

;***** Includes

.include "hw_spi.inc"
.include "hw_uart.inc"

;***** Subroutines

DELAY: clr     delcnt           ; Init Loop Counter
loop1: dec     delcnt
       nop
       brne    loop1
       dec     dly
       brne    loop1
       ret
```

Listing 5-20
Continued

```
;***** Main

RESET: ldi     temp, LOW(RAMEND)
       out     SPL, temp          ; Initialize SPL
       ldi     temp, HIGH(RAMEND)
       out     SPH, temp          ; Initialize SPH
       sbi     DDRB, PB0          ; PB0 output
       rcall   uart_init          ; Initialize UART
       rcall   spi_setup
       ldi     period, $FF        ; Initalize period
                                  ; (FF ok | 60 for error)

       ;ldi    spi_data, BL2_PROT ; Block Write Protection
                                  ; from $100 to $1FF
       ldi     spi_data, NO_BL_PROT ; No Block Write Protection
       rcall   write_en
       rcall   block_protection   ; Setup Status Register Bits in 25C04
       rcall   wait_for_spi_ready

       ldi     spi_addr, EE_addr
       ldi     spi_data, EE_data
       rcall   write_en
       rcall   spi_write          ; Store Byte in 25C04
       rcall   wait_for_spi_ready

       clr     spi_data

       ldi     spi_addr, EE_addr
       rcall   spi_read           ; Read back stored byte

       ;mov    temp, spi_data
       ;rcall  putc               ; Send it to terminal for test

       ldi     temp, EE_data
       cpse    spi_data, temp     ; Compare read back byte
                                  ; with byte written

       ldi     period, $60

loop:  sbi     PORTB, PB0
       mov     dly, period        ; Initialize delay controlled by period
       rcall   DELAY
       cbi     PORTB, PB0
       mov     dly, period        ; Initialize delay controlled by period
       rcall   DELAY
       rjmp    loop               ; Repeat forever
```

Listing 5-20
Continued

All communication must begin with a valid START condition and conclude with a STOP condition, and it must be acknowledged by the receiver with an ACKnowledge condition.

In addition, since the I²C bus is designed to support different devices such as EEPROM, RAM, and analog-to-digital and digital-to-analog converters, a device type identifier must follow the START condition. Figure 5-23 shows the required connections in a typical I²C bus network. The SDA and SCL line are pulled up to V_{CC} and connect the members of the network. In an I²C bus network, several masters can be connected to several slaves (multimaster system).

Before we look at to the I²C bus program example, some important and often-used terms should be explained. Table 5-4 lists some of these terms.

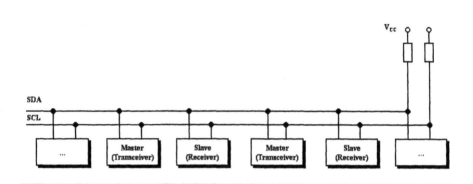

Figure 5-23
I²C bus network.

Table 5-4
Some I²C bus definitions.

Term	Meaning
WORD	8 data bits
PAGE	16 sequential memory locations
	(may be programmed during a Page Write programming cycle)
PAGE BLOCK	2048 bits organized into 16 pages of addressable memory
MASTER	Any I²C device controlling the transfer of data (a microcontroller, for example)
SLAVE	Device being controlled
TRANSMITTER	Device currently sending data on the bus (master or slave)
RECEIVER	Device currently receiving data on the bus (master or slave)
TRANSCEIVER	Device containing transmitter and receiver

For our I²C bus program example, we connect an I²C bus EEPROM to two I/O lines of an AVR microcontroller. Because of the required memory write and read operations, EEPROMs, and memories in general, are also well suited as example devices for this type of serial interface.

Each I²C bus device will be identified by a device type identifier contained in the slave address. Figure 5-24 shows the details of an EEPROM's slave address. The device type identifier is 1010_B in this case.

A further entry in this slave address is the device address. Device address pins A2, A1, and A0 are connected to V_{CC} or GND to configure the EEPROM chip address. Table 5-5 shows the active pins across the NM24Cxx device family.

As shown in Table 5-5, the EEPROMs on the I²C bus may be configured in any manner required; the total memory addressed in an I²C bus network cannot exceed 16K bits (16,384 bits). EEPROM memory addressing is controlled by two different methods:

(1) Hardware configuring the A2, A1, and A0 pins (Device Address pins) with pull-up or pull-down resistors. All unused pins (marked with x in Table 5-5) must be tied to GND.
(2) Software addressing the required Page Block within the device memory array (as sent in the Slave Address string).

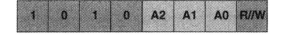

Figure 5-24
Slave address.

Table 5-5
Device addressing for I²C bus EEPROMs.

Device	A2	A1	A0	Memory Capacity
NM24C02	addr	addr	addr	2 K
NM24C04	addr	addr	x	4 K
NM24C08	addr	x	x	8 K
NM24C16	x	x	x	16 K

Addressing an EEPROM memory location involves sending a command string with the following information:

```
[DEVICE TYPE]-[DEVICE ADDRESS]-[PAGE BLOCK ADDRESS]-
[BYTE ADDRESS]
```

In our I²C bus program example, we use the NM24C16 EEPROM. Because of its memory capacity of 16K bits, no hardware configuration is required. Therefore, pins A2, A1, and A0 must be tied to GND.

The bits A2, A1, and A0 used in the slave address refer to an internal PAGE BLOCK memory segment. The last bit of the slave address defines whether a write or read condition is requested by the master. An LSB set to Hi indicates that a read operation is to be executed, whereas an LSB set to Lo initiates the write mode.

Two basic functions for data transmission over the I²C bus are Byte Write and Random Read operations. To make the EEPROM access more effective, further access modes such as Page Write, Current Address Read, and Sequential Read are defined.

In our I²C bus program example, we concentrate on the basic function to be demonstrated. Based on the described operations, an implementation of the enhanced functions is not so painful.

Figure 5-25 shows the bit sequences for Byte Write and Random Read operations. All commands are preceded by the START (S) condition, which is a Hi-to-Lo transition of SDA when SCL is Hi. Each I²C bus device, and, thus, the EEPROM as well, continuously monitors the SDA and SCL lines for a valid start condition and will not respond to any command until this condition has been met.

The first byte sent after a start condition is the slave address indicating a following write access to an addressed memory page. The transmitting device will release the bus after transmitting 8 bits. During the ninth clock cycle, the receiver will pull the SDA line to Lo to ACKnowledge (A) that it received the 8 bits of data. This acknowledge mechanism is a software convention used to

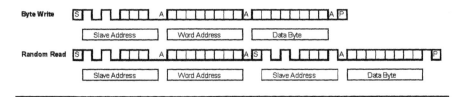

Figure 5-25
Byte Write and Random Read operations.

indicate successful data transfers. The second byte addresses the memory location in the preaddressed memory page for a following write or read access. After the last bit sent, the ACK is tested again.

In a Byte Write operation, the data byte that will be written in the addressed memory location is sent as the third byte. After the last bit sent, the ACK is tested again.

In a Random Read operation, a further start condition is sent immediately after the tested ACK. The first byte sent after this new start condition is the slave address indicating a following read access to the addressed memory location. After the last bit sent, the ACK is tested again. Next, the bits of the addressed memory location can be read from EEPROM.

In the read mode, the I²C bus slave will transmit 8 bits of data, release the SDA line, and monitor the line for an ACK from the master. If an acknowledge is detected and no stop condition is generated by the master, the slave will continue to transmit data. If an acknowledge is not detected, the slave will terminate further data transmissions and await the stop condition to return to the standby power mode.

All data transmissions are terminated by a STOP (P) condition, which is a Lo-to-Hi transition of SDA when SCL is Hi. The stop condition is also used by the NM24Cxx family to place the device in the standby power mode.

The basic functions for data transmission over an I²C bus are defined in the file i2c.inc shown in Listing 5-21. This include file is compatible with all AVR microcontrollers in principle. The program uses three levels of the stack space itself. Therefore, no stack space is free for a calling program running on an AT90S1200.

```
;******************************************************************
;* File Name      :i2c.inc
;* Title          :Software I2C
;* Author         :C.Kuehnel
;* Date           :10/05/97
;* Version        :1.0
;* Target MCU      :AT90S8515
;*
;* DESCRIPTION
;* Simple software I2C driver for AVR processors. Compatible with all AVR
;* processors: AT90S1200, AT90S8515, etc..
;******************************************************************
```

Listing 5-21
I²C Include file (i2c.inc).   (*Continued*)

```
; Primary routines:
; -----------------
; i2c_setup: Call to setup I/O bits for I2C. Must call prior to using the
;            following routines.
; i2c_start  Call for start condition
; i2c_stop   Call for stop condition
; i2c_write: Call to write I2C device with byte to send in i2c_buffer.
; i2c_read:  Call to read I2C device, byte returned in i2c_buffer.
;
; Registers used by software uart
; define in calling program as needed (must be >= r16)
; -------------------------------------------------------------------------
;       bitcntr             ; bit counter
;       i2c_del             ; delay loop variable
;       i2c_buffer          ; data byte transmit/received by I2C
;       bus_check           ; check for bus idle
;
; Equates defining software uart, define as needed
; -------------------------------------------------------------------------
.equ    i2c_delay   = 4     ; wait 4.75 us at 4 MHz
.equ    i2c_port    = PORTD ; port supporting i2c
.equ    i2c_pins    = PIND  ; port pins supporting i2c
.equ    i2c_ddr     = DDRD  ; data direction register for port
.equ    sda         = 6     ; bit within port for serial data in
.equ    scl         = 7     ; bit within port for serial data out
;
; Start and stop condition need setup resp. hold times of 4.7 us minimum.
; Clock Lo and Hi period must be 4.7 us minimum too.
; Required delay is build by a down-counter loop initialized with 4 for 4
; MHz clock frequency.
;
i2c_wait: ldi   i2c_del, i2c_delay    ; (1) + (3)
_i2c_wait_yet:  dec   i2c_del         ; (1)
                brne  _i2c_wait_yet    ; (1/2)
                ret                    ; (4)
;
; Set SDA and SCL to Hi level and switches SDA in a passive state
; for multi-master operation afterwards.
;
i2c_setup:      sbi   i2c_ddr, scl    ; set SCL to output
                sbi   i2c_ddr, sda    ; set SDA to output
                sbi   i2c_port,scl    ; set SCL Hi
                sbi   i2c_port,sda    ; set SDA Hi
                cbi   i2c_ddr, scl    ; set SCL to input (pull-up active)
                cbi   i2c_ddr, sda    ; set SDA to input (pull-up active)
                rcall i2c_wait
                ret
```

Listing 5-21
Continued

```
;
; Sends an I2C Start Condition to the I2C bus after check and wait for idle
; bus, means SDA = SCL = Hi.
;
i2c_start:      sbi     i2c_port,scl    ; set SCL Hi
                sbi     i2c_port,sda    ; set SDA Hi
                cbi     i2c_ddr, scl    ; set SCL to input (pull-up active)
                cbi     i2c_ddr, sda    ; set SDA to input (pull-up active)
                in      bus_check, i2c_pins          ; check I2C bus
                andi    bus_check, (1<<SDA)|(1<<SCL) ; mask SDA and SCL
                ldi     temp, (1<<SDA)|(1<<SCL)
                cp      bus_check, temp ; test for SDA=Hi and SCL=Hi
                brne    i2c_start
                sbi     i2c_ddr, scl    ; set SCL to output
                sbi     i2c_ddr, sda    ; set SDA to output
                cbi     i2c_port, sda   ; set SDA to Lo
                rcall   i2c_wait
                cbi     i2c_port, scl   ; set SCL to Lo
                rcall   i2c_wait
                ret


i2c_stop:sbi    i2c_port, scl           ; set SCL to Hi
                rcall   i2c_wait
                sbi     i2c_port, sda    ; set SDA to Hi
                cbi     i2c_ddr, scl     ; set SCL to input (pull-up active)
                cbi     i2c_ddr, sda     ; set SDA to input (pull-up active)
                rcall   i2c_wait
                ret
;
; Queries for I2C Achknowledge (SDA=Lo)
; T Flag in SREG serves as an error flag
; it is set if there was NAK and cleared if there was an ACK
;
i2c_ack:        clt                     ; clear error flag
                sbi     i2c_port,sda    ; set SDA Hi
                cbi     i2c_ddr, sda    ; set SDA to input (pull-up active)
                sbi     i2c_port, scl   ; set SCL to Hi
                rcall   i2c_wait
                sbic    i2c_pins, sda   ; reads SDA
                set                     ; set error flag when NAK
                cbi     i2c_port, scl   ; set SCL to Lo
                sbi     i2c_port, sda   ; set SDA to Hi
                rcall   i2c_wait
                ret
;
; Writes eight bits to I2C bus and queries the acknoledge bit from receiver
```

Listing 5-21
Continued

(Continued)

```
;
i2c_write:      sbi     i2c_ddr, sda        ; set SDA to output
                cbi     i2c_port, sda       ; set SDA to Lo
                ldi     bitcntr, 8          ; eight bits to write
_wrt_bit:       sbrc    i2c_buffer, 7       ; if MSB = Hi then
                sbi     i2c_port, sda       ; set SDA to Hi
                nop
                sbi     i2c_port, scl       ; set SCL to Hi
                rcall   i2c_wait
                cbi     i2c_port, scl       ; set SCL to Lo
                cbi     i2c_port, sda       ; set SDA to Lo
                rcall   i2c_wait
                lsl     i2c_buffer          ; shift i2c_buffer one bit left
                dec     bitcntr
                brne    _wrt_bit
                rcall   i2c_ack             ; query acknowledge
                ret
;
; Reads eight bits from I2C bus
;
i2c_read:       ldi     bitcntr, 8          ; eight bits to read
                clr     i2c_buffer          ; clear I2C buffer
                clc                         ; clear carry flag
                cbi     i2c_ddr, sda        ; set SDA to input
_rd_bit:        sbi     i2c_port, scl       ; set SCL to Hi
                sbic    i2c_pins, sda       ; if SDA = Hi then
                sec                         ; set carry flag
                rol     i2c_buffer          ; rotate i2c_buffer one bit left
                rcall   i2c_wait
                cbi     i2c_port, scl       ; set SCL to Lo
                rcall   i2c_wait
                dec     bitcntr
                brne    _rd_bit
                sbi     i2c_port, sda       ; set SDA to Hi
                sbi     i2c_ddr, sda        ; set SDA to output
                rcall   i2c_wait
                sbi     i2c_port, scl       ; set SCL to Hi
                rcall   i2c_wait
                cbi     i2c_port, scl       ; set SCL to Lo
                ret
```

Listing 5-21
Continued

As primary subroutines, i2c_setup, i2c_start, i2c_stop, i2c_write, and i2c_read are defined in this include file. Data exchange is handled by the register i2c_buffer. Secondary subroutines such as i2c_ack and i2c_wait are required from the subroutines i2c_write and i2c_read internally. In the case of an error during acknowledge, the T flag is set. This flag can be checked from error-handling routines.

Pin7 on PortD serves as clock output SCL, while Pin6 on PortD serves as bidirectional I/O line SDA.

The constant i2c_delay is responsible for timing. Minimum clock periods for SCL = Lo and SCL = Hi of 4.75 µs are demanded from I²C bus specifications. For a clock frequency of 4 MHz, calling the subroutine i2c_wait delays the execution of the next instruction for this length of time. The execution time of the single instructions in the subroutine i2c_wait, including the time for calling and return, is noted as comment in the source. Adaptions to higher clock frequencies are possible.

For testing the I²C bus data exchange with the EEPROM NM24C16, two small programs running on Atmel's evaluation board were written. Listing 5-22 shows a program that writes one predefined data byte to a predefined memory location in EEPROM.

```
;********************************************************************
;* File Name          :i2c_bw.asm
;* Title              :Software I2C Byte Write Test
;* Author             :C.Kuehnel
;* Date               :10/12/97
;* Version            :1.0
;* Target MCU         :AT90S8515
;*
;* DESCRIPTION
;* Simple software I2C driver for AVR processors. Compatible with all AVR
;* processors: AT90S1200, AT90S8515, etc..
;********************************************************************

;***** Directives
.device    at90s8515
.nolist
.include "8515def.inc"
.list
```

Listing 5-22
I²C byte write test (i2c_bw.asm).   (*Continued*)

```
; Registers used by I2C
; define in calling program as needed (must be >= r16)
; ----------------------------------------------------------------------
.def    bitcntr     = r20   ; bit counter
.def    i2c_del     = r21   ; delay loop variable
.def    i2c_buffer  = r22   ; async serial data byte transmit/received
.def    bus_check   = r23   ; check for bus idle
;
; Main Program Register variables
; ----------------------------------------------------------------------
.def    temp        = r16
.def    byte        = r17
.def    dly         = r18
.def    delcnt      = r19   ; Loop Counter

.equ    device_id   = $A    ; Device Identification for 24C16 EEPROM
.equ    page_addr   = 1     ; Page address
.equ    word_addr   = 0     ; Memory location
.equ    ee_data     = $A5   ; Test pattern


;***** Interrupt vector table

            rjmp    RESET       ; Reset handle
            reti                ; External Interrupt0 handle
            reti                ; External Interrupt1 handle
            reti                ; T/C1 Capture Event Interrupt handle
            reti                ; T/C1 CompareA Interrupt handle
            reti                ; T/C1 CompareB Interrupt handle
            reti                ; T/C1 Overflow Interrupt handle
            reti                ; T/C0 Overflow Interrupt handle
            reti                ; SPI Transfer Complete Interrupt handle
            reti                ; UART Rx Complete Interrupt handle
            reti                ; UART Data Register Empty Interrupt handle
            reti                ; UART Tx Complete Interrupt handle
            reti                ; Analog Comparator Interrupt handle

;***** Includes

.include "i2c.inc"

;***** Subroutines
```

Listing 5-22
Continued

```
DELAY:    clr     delcnt          ; Init Loop Counter
loop1:    dec     delcnt
          nop
          brne    loop1
          dec     dly
          brne    loop1
          ret


;***** Main

RESET:    ldi     temp, LOW(RAMEND)
          out     SPL, temp           ; Initialize SPL
          ldi     temp, HIGH(RAMEND)
          out     SPH, temp           ; Initialize SPH
          ser     temp
          out     DDRB, temp
          out     PORTB,temp          ; Setup PortB to output

          rcall   i2c_setup           ; Initialize UART

          rcall   i2c_start           ; I2C Start Condition
          ldi     i2c_buffer, (device_id<<4)|(page_addr<<1)
                                      ; Slave address =b10100010
          rcall   i2c_write           ; Write Slave address
          ldi     i2c_buffer, word_addr
          rcall   i2c_write           ; Write Word address
          ldi     i2c_buffer, ee_data
          rcall   i2c_write           ; Write Data byte
          rcall   i2c_stop            ; I2C Stop Condition

loop:     sbi     PORTB, PB0
          ldi     dly, $FF            ; Initialize delay of about 65 ms
          rcall   DELAY               ; Number of cycles = 1026 * del + 4
          cbi     PORTB, PB0
          ldi     dly, $FF            ; Initialize delay of about 65 ms
          rcall   DELAY               ; Number of cycles = 1026 * del + 4
          rjmp    loop                ; Repeat forever
```

Listing 5-22
Continued

To simplify matters, some parameters were defined by directives as constants:

- Device identifier for all EEPROMs of the NM24Cxx family `device_id = $A`
- Memory access to page 1 address 0 by `page_addr = 1` and `word_addr = 0` (arbitrarily)
- Data byte for storage `ee_data = $A5` (arbitrarily)

After initalizing the stack pointer and PortB for signalizing by LEDs, the single steps of a byte write sequence are noted. Please compare the subroutine calls with Figure 5-25 (top sequence). To keep it simple, an error handler checking the T flag was not implemented.

Following the byte write procedure, an endless loop blinks an LED connected to Pin0 of PortB. The blinking LED signals that the data byte is written.

To test the correctness, the written data byte must be read back. Listing 5-23 shows a program for reading the EEPROM NM24C16 over the I²C bus. The program `i2c_br.asm` is quite similar to `i2c_bw.asm`. The byte write sequence was replaced by the byte read sequence. Please compare the subroutine calls with Figure 5-25 (bottom sequence).

```
;*****************************************************************
;* File Name          :i2c_br.asm
;* Title              :Software I2C Byte Read Test
;* Author             :C.Kuehnel
;* Date               :10/12/97
;* Version            :1.0
;* Target MCU         :AT90S8515
;*
;* DESCRIPTION
;* Simple software I2C driver for AVR processors. Compatible with all AVR
;* processors: AT90S1200, AT90S8515, etc..
;*****************************************************************

;***** Directives
.device    at90s8515
.nolist
.include "8515def.inc"
.list
```

Listing 5-23
I²C byte read test (i2c_br.asm).

```
; Registers used by I2C
; define in calling program as needed (must be >= r16)
; ------------------------------------------------------------------------
.def      bitcntr      = r20    ; bit counter
.def      i2c_del      = r21    ; delay loop variable
.def      i2c_buffer   = r22    ; async serial data byte transmit/received
.def      bus_check    = r23    ; check for bus idle
;
; Main Program Register variables
; ------------------------------------------------------------------------
.def      temp         = r16
.def      byte         = r17
.def      dly          = r18
.def      delcnt       = r19    ; Loop Counter

.equ      device_id    = $A     ; Device Identification for 24C16 EEPROM
.equ      page_addr    = 1      ; Page address
.equ      word_addr    = 0      ; Memory location

;***** Interrupt vector table

          rjmp      RESET       ; Reset handle
          reti                  ; External Interrupt0 handle
          reti                  ; External Interrupt1 handle
          reti                  ; T/C1 Capture Event Interrupt handle
          reti                  ; T/C1 CompareA Interrupt handle
          reti                  ; T/C1 CompareB Interrupt handle
          reti                  ; T/C1 Overflow Interrupt handle
          reti                  ; T/C0 Overflow Interrupt handle
          reti                  ; SPI Transfer Complete Interrupt handle
          reti                  ; UART Rx Complete Interrupt handle
          reti                  ; UART Data Register Empty Interrupt handle
          reti                  ; UART Tx Complete Interrupt handle
          reti                  ; Analog Comparator Interrupt handle

;***** Includes

.include "i2c.inc"

;***** Subroutines

DELAY:    clr       delcnt      ; Init Loop Counter
loop1:    dec       delcnt
```

Listing 5-23
Continued

(*Continued*)

```
            nop
            brne    loop1
            dec     dly
            brne    loop1
            ret


;***** Main

RESET:      ldi     temp, LOW(RAMEND)
            out     SPL, temp               ; Initialize SPL
            ldi     temp, HIGH(RAMEND)
            out     SPH, temp               ; Initialize SPH
            ser     temp
            out     DDRB, temp
            out     PORTB,temp              ; Setup PortB to output


            rcall   i2c_setup               ; Initialize UART


            rcall   i2c_start               ; I2C Start Condition
            ldi     i2c_buffer, (device_id<<4)|(page_addr<<1)
                                            ; Slave address =b10100010
            rcall   i2c_write               ; Write Slave address
            ldi     i2c_buffer, word_addr
            rcall   i2c_write               ; Write Word address
            rcall   i2c_start               ; I2C Start Condition
            ldi     i2c_buffer, (device_id<<4)|(page_addr<<1) | 1
                                            ; Slave address =b10100011
            rcall   i2c_write               ; Write Slave address
            rcall   i2c_read                ; Read Data byte
            rcall   i2c_stop                ; I2C Stop Condition


            out     PORTB, i2c_buffer       ; Output Read byte

loop:       sbi     PORTB, PB0
            ldi     dly, $FF                ; Initialize delay of about 65 ms
            rcall   DELAY                   ; Number of cycles = 1026 * del + 4
            cbi     PORTB, PB0
            ldi     dly, $FF                ; Initialize delay of about 65 ms
            rcall   DELAY                   ; Number of cycles = 1026 * del + 4
            rjmp    loop                    ; Repeat forever
```

Listing 5-23
Continued

After reading the data byte from the same location (page = 1, address = 0), the read data byte is output to PortB and visualized by the connected LEDs. Following the byte read procedure and output, an endless loop blinks the LED connected to Pin0. The blinking LED signals that the data byte has been read.

5.2 Example Program in C

Programming the AVR microcontrollers in C will be demonstrated with a simple interrupt-driven event counter. The External Interrupt 0 reacts to events detected at Pin PD2. Listing 5-24 shows the C source text written for AT90S8515.

The program example itself is easy to understand even for non-C programmers. The interrupt handler for the external interrupt toggles a flag and I/O pin PD0. The variable event_count is incremented at each access to this handle. To visualize the count, it is written to PortB and sent via serial interface to a terminal, for example.

In the main part, there are only two initializations of interrupts before the program enters an endless loop.

The generated listing file shows the conversion of C to assembler code. It can be very interesting to have a look at this file to get an impression of how the compiler works. Listing 5-25 shows the result of the compiling process.

Because it is not possible to generate executable code with the demo of the IAR Embedded Workbench, this program example is explained in the simulation environment.

First, we look at Figure 5-26, which shows the IAR Embedded Workbench with opened windows for project, source text, and messages from the compilation process. If the compiling process was successful (that is, no errors), the simulation process can be started by pressing the C-SPY button in the toolbar (Figure 5-27).

All usual C-SPY simulation tasks are possible within the C-SPY simulator. Figure 5-28 shows the user interface of the C-SPY simulator. The user interface can be arranged for optimal simulation conditions. In the Source window, we can inspect the program execution step-by-step or, as shown in Figure 5-28, by setting breakpoints. The breakpoint is set inside the interrupt handle so the program execution will stop if simulated internal interrupt occurs. The Report window displays the break and other actions.

In the Register window, only the interesting resources are displayed. Both the I/O ports used are important in our example. PortB shows the event count and PortD the toggling I/O line (PD0). The serial output is visualized in the terminal I/O window. All characters sent to the terminal are displayed in this window.

```
/*******************************************************/
// File Name        : ext_interrupt.c
// Title            : External Interrupt Test
// Author           : C.Kuehnel
// Date             : 11/01/97
// Target MCU        : AT90S8515
/*******************************************************/

/* enable use of extended keywords */
#pragma language=extended

/* include sfr definitions for IO registers */
#include <io8414.h>          // must be replaced by updated header file
#include <ina90.h>
#include <stdio.h>

#define TRUE 1
#define FALSE 0

interrupt [INT0_vect] void External_interrupt(void)
{
  static char flag;
  static int event_count;

  if (flag)
    flag = FALSE;
  else
    flag = TRUE;
  PORTD = flag & 0x01;      // Toggle PD0
  PORTB= ++event_count;      // Increment Event Counter and write to PortB
  putchar(event_count);      // Send Event Counter Contents to Terminal
}

void main(void)
{
    GIMSK |= 0x40;          // INT0 enabled
    _SEI();                 // Enable interrupts
    while(TRUE);
}
```

Listing 5-24
Interrupt-driven event counter (`ext_interrupt.c`).

```
################################################################################
#                                                                              #
# IAR AT90S C-Compiler V1.10A/386 [demo]                                       #
# Front End V4.20N                              01/Nov/97 14:30:58             #
# Global Optimizer V1.05D                                                      #
#                                                                              #
#     Target option = Max 64 KB data + 8 KB code (8414...)                     #
#     Memory model  = tiny                                                     #
#     Source file   = d:\programme\avr\c_examples\ext_interrupt.c             #
#     List file     = d:\programme\avr\c_examples\release\list\ext_interrupt.1st #
#     Object file   = d:\programme\avr\c_examples\release\obj\ext_interrupt.r90 #
#     Command line  = -v1 -mt -OD:\Programme\AVR\C_Examples\Release\Obj\       #
#                     -e -K -gA -z9 -RCODE -r0                                 #
#                     -LD:\Programme\AVR\C_Examples\Release\List\ -q -t8       #
#                     -X -ID:\PROGRAMME\EWA90\A90\inc\                         #
#                     D:\Programme\AVR\C_Examples\ext_interrupt.c              #
#                                                                              #
#                                     (c) Copyright IAR Systems 1996 #
################################################################################
      1          /*****************************************************/
      2          // File Name  : ext_interrupt.c
      3          // Title      : External Interrupt Test
      4          // Author     : C.Kuehnel
      5          // Date       : 11/01/97
      6          // Target MCU : AT90S8515
      7          /*****************************************************/
      8
      9          /* enable use of extended keywords */
     10          #pragma language=extended
     11
     12          /* include sfr definitions for IO registers */
     13          #include <io8414.h>
     14          #include <ina90.h>
     15          #include <stdio.h>
     16
     17          #define TRUE 1
     18          #define FALSE 0
     19
     20
     21          interrupt [INT0_vect] void External_interrupt(void)
     22          {
\ 000000   ....               RCALL   ?INT_PROLOGUE_FIRST_L09
     23            static char flag;
     24            static int event_count;
     25
     26          if (flag)
\ 000002   ....               LDI     R30,?0002
\ 000004   FF27               CLR     R31
```

Listing 5-25
List file ext_interrupt.1st.   (Continued)

```
\ 000006   0081            LDD    R16,Z+0
\ 000008   0023            TST    R16
\ 00000A   11F0            BREQ   ?0005
 27            flag = FALSE;
\ 00000C   00E0            LDI    R16,0
\ 00000E   01C0            RJMP   ?0010
 28          else
 29            flag = TRUE;
\ 000010   01E0            LDI    R16,1
\ 000012   0083            STD    Z+0,R16
 30          PORTD = flag & 0x01;    // Toggle PD0
\ 000014   0170            ANDI   R16,LOW(1)
\ 000016   02BB            OUT    LOW(18),R16
 31          PORTB= ++event_count; // Increment Event Counter and write to PortB
\ 000018   ....            LDI    R30,?0003
\ 00001A   0081            LDD    R16,Z+0
\ 00001C   1181            LDD    R17,Z+1
\ 00001E   0F5F            SUBI   R16,LOW(255)
\ 000020   1F4F            SBCI   R17,LOW(255)
\ 000022   0083            STD    Z+0,R16
\ 000024   1183            STD    Z+1,R17
\ 000026   08BB            OUT    LOW(24),R16
 32          putchar(event_count); // Send Event Counter Contents to Terminal
\ 000028   .... RCALL putchar
 33          }
\ 00002A   E0E0            LDI    R30,0
\ 00002C   ....            RJMP   ?INT_EPILOGUE0_L09
 34
 35
 36          void main(void)
 37          {
 38            GIMSK |= 0x40;    // INT0 enabled
\ 00002E   0BB7            IN     R16,LOW(59)
\ 000030   0064            ORI    R16,LOW(64)
\ 000032   0BBF            OUT    LOW(59),R16
 39            _SEI();            // Enable interrupts
\ 000034   7894            SEI
 40            while(TRUE);
 41          }
\ 000036   FFCF            RJMP   ?0008

Errors: none
Warnings: none
Code size: 56
Constant size: 4
Static variable size: 3
```

Listing 5-25
Continued

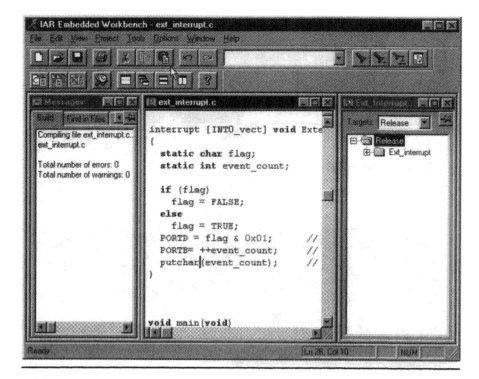

Figure 5-26
Project Ext_Interrupt in IAR Embedded Workbench.

Figure 5-27
Toolbar.

5.3 Example Programs in AVR BASIC

5.3.1 Microcontroller Test

The first example, an AVR BASIC program, is again a microcontroller test program. Its functionality is the same as that of the Assembler program described earlier.

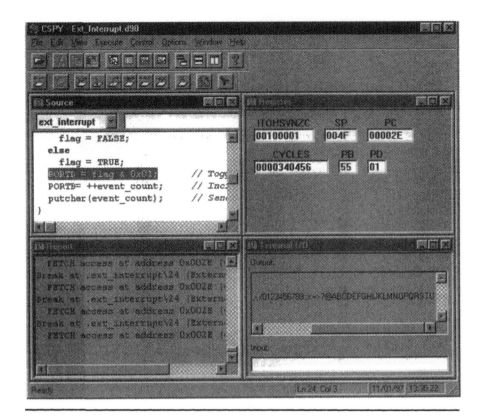

Figure 5-28
Simulation of Ext_Interrupt test in C-SPY simulator.

The Timer/Counter0 works as a free running timer. The Timer Overflow interrupt outputs a bit pattern to PortB, so a permanently changing bit pattern can be detected. The refresh rate depends on a reload value that can be manipulated by an external interrupt generated by a pressed key, for example.

Listing 5-26 shows this microcontroller test program written in AVR BASIC.

The register names are all identical to the Atmel definitions, so the source text is easy to read. It is interesting to compare the AVR BASIC source text with the generated list file (Listing 5-27).

5.3.2 Pulse-Width Modulation and Serial Communication for the AT90S8515

The example program, Listing 5-28, demonstrates the use of the internal resources in an application with the AT90S8515 microcontroller. It shows an

```
;  ********************************************************************
' File Name              :MC_TEST.BAS
' Title                  :Microcontroller Test
' Date                   :30/08/97
' Version                :1.0
' Target MCU             :AT90S1200
'
' DESCRIPTION
' Test of Timer0 Overflow and External Interrupt on AVR Evaluation Board
'  ********************************************************************
'
' Symbol definitions
'
Symbol temp = R16                       ' Low Bank OK
Symbol byte = R17                       ' Low Bank OK
Symbol reload = R18                     ' Low Bank OK
'
' AVR Status Register Symbolic Names
'
Symbol CARRY           = SREG.0
Symbol ZERO            = SREG.1
Symbol HALF_CARRY      = SREG.5
Symbol BIT_STORAGE     = SREG.6
Symbol GIE             = SREG.7
'
' Interrupt Vector Table
'
        Goto Start      ' Reset Handle
        Goto Ex_Int0    ' External Interrupt0 Handle
        Goto OVF0       ' Overflow0 Interrupt Handle
        Reti
'
' Interrupt Handlers
'
Ex_Int0: Swap reload
        Reti

OVF0:   PortB = byte
        rol byte
        TCNT0 = reload
        Reti
```

Listing 5-26
AVR BASIC microcontroller test (mc_test-bas).   (*Continued*)

```
'
' Main
'

Start:    reload = $07              ' Initialize Reload Value


          carry = 1
          byte = $FE
          PORTB = byte              ' Initialize PortB


          DDRB = $FF


          TCCR0 = %00000101         ' Prescaler CK/1024 => 256 us


          TIMSK = %00000010         ' Enable Timer Interrupts


          temp = MCUCR              ' Interrupt on falling edge of INT0
          or temp, %00000010
          MCUCR = temp


          temp = GIMSK              ' External Interrupt Enable
          or temp, %01000000
          GIMSK = temp


          GIE = 1                   ' Global Enable Interrupts


loop:     goto loop
```

Listing 5-26
Continued

example program that generates a pulse-width-modulated output and echoes a received byte back to the transmitter.

Before generating a pulse-width-modulated output and communication over the UART, both peripherals must be initialized. The main program responsible for the serial communication polls the UART Receive Complete bit (RXC) until a character is received. The received character is mirrored to the UART data register to send back afterwards.

The instruction UDR = UDR seems a bit confusing at first. Remember that the register UDR is actually two physically separate registers sharing the same address in the I/O area.

```
   1: ' *********************************************************************
   2: ' File Name          :MC_TEST.BAS
   3: ' Title              :Microcontroller Test
   4: ' Date               :30/08/97
   5: ' Version            :1.0
   6: ' Target MCU         :AT90S1200
   7: '
   8: ' DESCRIPTION
   9: ' Test of Timer0 Overflow and External Interrupt on AVR Evaluation Board
  10: ' *********************************************************************
  11: '
  12: ' Symbol definitions
  13: '
  14: Symbol temp = R16                       ' Low Bank OK
  15: Symbol byte = R17                       ' Low Bank OK
  16: Symbol reload = R18                     ' Low Bank OK
  17:
  18: '
  19: 'AVR Status Register Symbolic Names
  20: '
  21: Symbol CARRY = SREG.0
  22: Symbol ZERO = SREG.1
  23: Symbol HALF_CARRY = SREG.5
  24: Symbol BIT_STORAGE = SREG.6
  25: Symbol GIE = SREG.7
  26:
  27: '
  28: ' Interrupt Vector Table
  29: '
  30:     Goto Start          ' Reset Handle
0000:C009 RJMP 0x009 [30]
  31:     Goto Ex_Int0        ' External Interrupt0 Handle
0001:C002 RJMP 0x002 [31]
  32:     Goto OVF0           ' Overflow0 Interrupt Handle
0002:C003 RJMP 0x003 [32]
  33:     Reti
0003:9518 RETI [33]
  34: '
  35: ' Interrupt Handlers
```

Listing 5-27
List file mc_test.lst generated from mc_test.bas.   (*Continued*)

```
 36: '
 37: Ex_Int0:Swap reload
0004:9522 SWAP R18 [37]
 38:      Reti
0005:9518 RETI [38]
 39: OVF0:      PortB = byte
0006:BB18 OUT 0x18,R17 [39]
 40:      rol byte
0007:1F11 ADC R17,R17 [40]
 41:      TCNT0 = reload
0008:BF22 OUT 0x32,R18 [41]
 42:      Reti
0009:9518 RETI [42]
 43:
 44: '
 45: ' Main
 46: '
 47: Start:       reload = $07      ' Initialize Reload Value
000A:E027 LDI R18,0x07 [47]
 48:
 49:      carry = 1
000B:9408 SEC [49]
 50:      byte = $FE
000C:EF1E LDI R17,0xFE [50]
 51:      PORTB = byte             ' Initialize PortB
000D:BB18 OUT 0x18,R17 [51]
 52:
 53:      DDRB = $FF
000E:EFFF LDI R31,0xFF [53]
000F:BBF7 OUT 0x17,R31 [53]
 54:
 55:      TCCR0 = %00000101        ' Prescaler CK/1024 => 256 us
0010:E0F5 LDI R31,0x05 [55]
0011:BFF3 OUT 0x33,R31 [55]
 56:
 57:      TIMSK = %00000010        ' Enable Timer Interrupts
0012:E0F2 LDI R31,0x02 [57]
0013:BFF9 OUT 0x39,R31 [57]
 58:
 59:      temp = MCUCR             ' Interrupt on falling edge of INT0
0014:B705 IN R16,0x35 [59]
 60:      or temp, %00000010
```

Listing 5-27
Continued

```
0015:6002 ORI R16,0x02 [60]
  61:     MCUCR = temp
0016:BF05 OUT 0x35,R16 [61]
  62:
  63:     temp = GIMSK          ' External Interrupt Enable
0017:B70B IN R16,0x3B [63]
  64:     or temp, %01000000
0018:6400 ORI R16,0x40 [64]
  65:     GIMSK = temp
0019:BF0B OUT 0x3B,R16 [65]
  66:
  67:     GIE = 1               ' Global Enable Interrupts
001A:9478 SEI [67]
  68:
  69: loop:    goto loop
001B:CFFF RJMP -0x001 [69]
  70:
  71:
  72:
```

```
* SYMBOL TABLE *
WREG                    REG VAR     001F
TEMP                    REG VAR     0010
BYTE                    REG VAR     0011
RELOAD                  REG VAR     0012
CARRY                   I/O REG BIT 003F.0
ZERO                    I/O REG BIT 003F.1
HALF_CARRY              I/O REG BIT 003F.5
BIT_STORAGE             I/O REG BIT 003F.6
GIE                     I/O REG BIT 003F.7

* LABELS *
EX_INT0                 0004
OVF0                    0006
START                   000A
LOOP                    001B
U                       001C
```

```
Program Memory used : 28 words
Internal EEPROM used: 0 bytes
Errors: 0
```

Listing 5-27
Continued

Example Programs   271

```basic
'
' AVR Basic AT90S8515 Test Program
' http://avrbasic.com
'
' All I/O pins set to output low(0), except:
' Hardware UART initialized for 9600 @11.0592
' PWM A [pin 15] set to 1/3 VCC
' PWM B [pin 29] set to 2/3 VCC
'
' compiler build >= 1.0.0.102
'
DEVICE 8515                     'Select AT90S8515 as Target AVR
'
' Define some UART Symbolic Names
'
Symbol UART_TX_Empty = USR.5    '1 When Ok to write to UDR
Symbol UART_RX_Char = USR.7     '1 When Character available
'
' Global Init
'
SPL = $FF                       'In case we use Gosub, we need Stack Pointer
'
' Set all ports to output
'
DDRA = $FF; DDRB = $FF; DDRC = $FF; DDRD = $FF
'
' Init Both PWM Channels
'
TCCR1B = $01                    'Start Timer 1
TCCR1A = %10100001              'Enable both PWM Channels, 8 bit
OCR1AL = 255 / 3                'Set PWM A to 1/3 of VCC
OCR1BL = 255 * 2 / 3            'Set PWM B to 2/3 of VCC
'
' Init UART
'
UBRR = 71                       '9600 Baud at 11.0592 Clock
UCR.4 = 1                       'Enable Receiver
UCR.3 = 1                       'Enable Transmitter
'
' Really simple Character ECHO Program
'
Main:
 Wait UART_RX_Char              'Wait till input
 UDR = UDR                      'Get Character/Put Character
 Goto Main                      'Loop Forever
```

Listing 5-28
PWM and RS232 with AT90S8515 (test8515.bas).

Appendix A
Part Numbering System

The AVR microcontroller family has an uniform part numbering system for all members of the family. Figure A-1 shows the details of the part numbering system. Table A-1 explains the coding for the amount of EEPROM and SRAM data memory.

The CPU and peripheral resources of the first members of the AVR microcontroller family are listed in Table A-2.

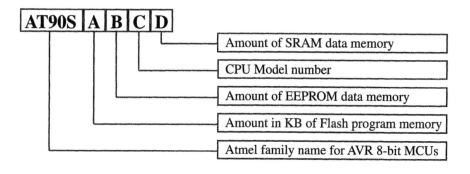

Figure A-1
Part numbering system for AVR microcontrollers.

Table A-1
Coding of amount of EEPROM and SRAM data memory.

Code	0	1	2	3	4	5	6	7	8	9	A	B	...
Memory Amount	0	32	64	128	256	512	1K	2K	4K	8K	16K	32K	...

Table A-2

First AVR family members and their CPU resources.

Family Member	Flash [KB]	EEPROM [Byte]	CPU Model	SRAM [Byte]	Counter Timer	UART	ADC	Pins	Available
AT90S1200	1	64	0	0	1	No	No	20	Q197
AT90S1220	1	64	2	0	1	No	No	8	Q397
AT90S2313	2	128	1	128	2	Yes	No	20	Q297
AT90S4414	4	256	1	256	3	Yes	No	40/44	Q397
AT90S4433	4	256	3	128	?	Yes	Yes	28	Q497
AT90S8515	8	512	1	512	3	Yes	No	40/44	Q297
AT90S68718	68	2048	1	4096					

Appendix B
Pin Configurations

The pinout of the first devices in the AVR microcontroller family is shown in Figures B-1 to B-3.

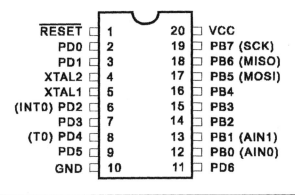

Figure B-1
Pin diagram for AT90S1200.

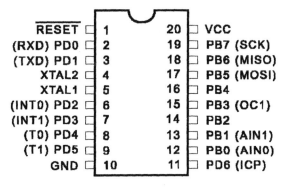

Figure B-2
Pin diagram for AT90S2313.

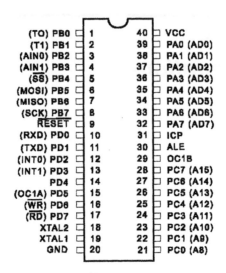

(T0) PB0	1		40	VCC
(T1) PB1	2		39	PA0 (AD0)
(AIN0) PB2	3		38	PA1 (AD1)
(AIN1) PB3	4		37	PA2 (AD2)
($\overline{SS}$) PB4	5		36	PA3 (AD3)
(MOSI) PB5	6		35	PA4 (AD4)
(MISO) PB6	7		34	PA5 (AD5)
(SCK) PB7	8		33	PA6 (AD6)
$\overline{RESET}$	9		32	PA7 (AD7)
(RXD) PD0	10		31	ICP
(TXD) PD1	11		30	ALE
(INT0) PD2	12		29	OC1B
(INT1) PD3	13		28	PC7 (A15)
PD4	14		27	PC6 (A14)
(OC1A) PD5	15		26	PC5 (A13)
($\overline{WR}$) PD6	16		25	PC4 (A12)
($\overline{RD}$) PD7	17		24	PC3 (A11)
XTAL2	18		23	PC2 (A10)
XTAL1	19		22	PC1 (A9)
GND	20		21	PC0 (A8)

Figure B-3
Pin diagram for AT90S4414 and 8515.

Appendix C
Schematics of SIMMSTICK Modules

The schematics of the SIMMSTICK Modules DT103 and DT104 are reproduced here with permission of DonTronics. Both these modules serve as examples of hardware solutions around the AVR microcontrollers.

Both modules are part of the SIMMSTICK module family and can be used as prototyping boards or as programmers.

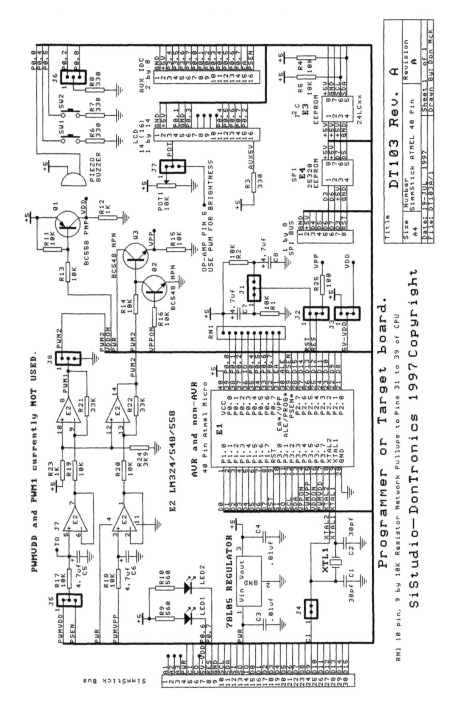

SIMMSTICK module DT104 for 20-pin AVR microcontrollers.

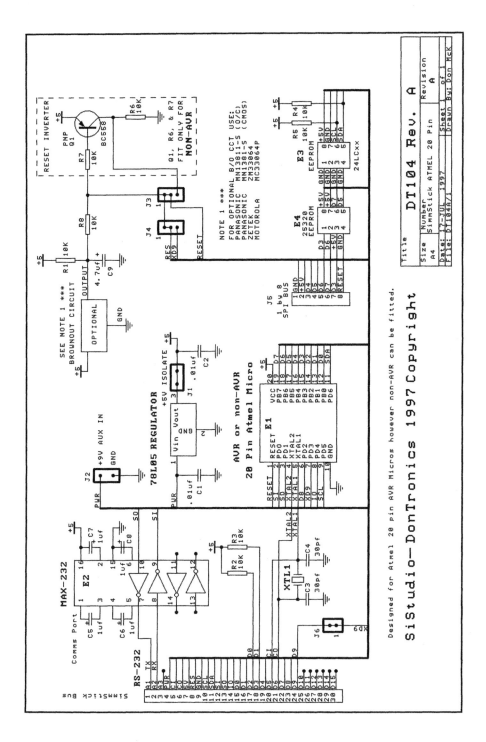

SIMMSTICK module DT103 for 40-pin AVR microcontrollers.

Appendix D
Register and Bit Definitions

For each device in the AVR microcontroller family, some definitions of registers, bits, and addresses are predefined in an include file xxxdef.inc. The characters xxxx are replaced by the part number of the microcontroller used. The following listing shows the contents of the file 8515def.inc as an example.

```
*************************************************************************
;* A P P L I C A T I O N   N O T E   F O R   T H E   A V R   F A M I L Y
;*
;* Number           :AVR000
;* File Name         :"8515def.inc"
;* Title             :Register/Bit Definitions for the AT90S8515
;* Date              :97.01.28
;* Version           :1.11
;* Support telephone :+47 72 88 87 20 (ATMEL Norway)
;* Support fax       :+47 72 88 87 18 (ATMEL Norway)
;* Support email     :avr@atmel.com
;* Target MCU        :AT90S8515
;*
;* DESCRIPTION
;* When including this file in the assembly program file, all I/O register
;* names and I/O register bit names appearing in the data book can be used.
;* In addition, the six registers forming the three data pointers X, Y, and
;* Z have been assigned names XL - ZH. Highest RAM address for Internal
;* SRAM is also defined.
;*
;* The Register names are represented by their hexadecimal address.
;*
;* The Register Bit names are represented by their bit number (0-7).
;*
;* Please observe the difference in using the bit names with instructions
;* such as "sbr"/"cbr" (set/clear bit in register) and "sbrs"/"sbrc"
;* (skip if bit in register set/cleared). The following example
```

```
;* illustrates this:
;*
;* in        r16,PORTB                ;read PORTB latch
;* sbr       r16,(1<<PB6)+(1<<PB5)    ;set PB6 and PB5 (use masks, not bit#)
;* out    PORTB,r16                   ;output to PORTB
;*
;* in        r16,TIFR                 ;read the Timer Interrupt Flag Register
;* sbrc      r16,TOV0                 ;test the overflow flag (use bit#)
;* rjmp      TOV0_is_set              ;jump if set
;* ...                                ;otherwise do something else
;************************************************************************

;***** I/O Register Definitions
.equ        SREG      =$3f
.equ        SPH       =$3e
.equ        SPL       =$3d
.equ        GIMSK     =$3b
.equ        TIMSK     =$39
.equ        TIFR      =$38
.equ        MCUCR     =$35
.equ        TCCR0     =$33
.equ        TCNT0     =$32
.equ        TCCR1A    =$2f
.equ        TCCR1B    =$2e
.equ        TCNT1H    =$2d
.equ        TCNT1L    =$2c
.equ        OCR1AH    =$2b
.equ        OCR1AL    =$2a
.equ        OCR1BH    =$29
.equ        OCR1BL    =$28
.equ        ICR1H     =$25
.equ        ICR1L     =$24
.equ        WDTCR     =$21
.equ        EEARH     =$1f
.equ        EEARL     =$1e
.equ        EEDR      =$1d
.equ        EECR      =$1c
.equ        PORTA     =$1b
.equ        DDRA      =$1a
.equ        PINA      =$19
.equ        PORTB     =$18
.equ        DDRB      =$17
.equ        PINB      =$16
.equ        PORTC     =$15
.equ        DDRC      =$14
.equ        PINC      =$13
.equ        PORTD     =$12
.equ        DDRD      =$11
.equ        PIND      =$10
.equ        SPDR      =$0f
```

```
.equ         SPSR        =$0e
.equ         SPCR        =$0d
.equ         UDR         =$0c
.equ         USR         =$0b
.equ         UCR         =$0a
.equ         UBRR        =$09
.equ         ACSR        =$08

;***** Bit Definitions
.equ         INT1        =7
.equ         INT0        =6
.equ         TOIE1       =7
.equ         OCIE1A      =6
.equ         OCIE1B      =5
.equ         TICIE       =3
.equ         TOIE0       =1
.equ         TOV1        =7
.equ         OCF1A       =6
.equ         OCF1B       =5
.equ         ICF1        =3
.equ         TOV0        =1
.equ         SRE         =7
.equ         SRW         =6
.equ         SE          =5
.equ         SM          =4
.equ         ISC11       =3
.equ         ISC10       =2
.equ         ISC01       =1
.equ         ISC00       =0
.equ         CS02        =2
.equ         CS01        =1
.equ         CS00        =0
.equ         COM1A1      =7
.equ         COM1A0      =6
.equ         COM1B1      =5
.equ         COM1B0      =4
.equ         PWM11       =1
.equ         PWM10       =0
.equ         ICNC1       =7
.equ         ICES1       =6
.equ         CTC1        =3
.equ         CS12        =2
.equ         CS11        =1
.equ         CS10        =0
.equ         WDE         =3
.equ         WDP2        =2
.equ         WDP1        =1
.equ         WDP0        =0
.equ         EEAR8       =0
.equ         EEWE        =1
```

```
.equ        EERE        =0
.equ        PA7         =7
.equ        PA6         =6
.equ        PA5         =5
.equ        PA4         =4
.equ        PA3         =3
.equ        PA2         =2
.equ        PA1         =1
.equ        PA0         =0
.equ        DDA7        =7
.equ        DDA6        =6
.equ        DDA5        =5
.equ        DDA4        =4
.equ        DDA3        =3
.equ        DDA2        =2
.equ        DDA1        =1
.equ        DDA0        =0
.equ        PINA7       =7
.equ        PINA6       =6
.equ        PINA5       =5
.equ        PINA4       =4
.equ        PINA3       =3
.equ        PINA2       =2
.equ        PINA1       =1
.equ        PINA0       =0
.equ        PB7         =7
.equ        PB6         =6
.equ        PB5         =5
.equ        PB4         =4
.equ        PB3         =3
.equ        PB2         =2
.equ        PB1         =1
.equ        PB0         =0
.equ        DDB7        =7
.equ        DDB6        =6
.equ        DDB5        =5
.equ        DDB4        =4
.equ        DDB3        =3
.equ        DDB2        =2
.equ        DDB1        =1
.equ        DDB0        =0
.equ        PINB7       =7
.equ        PINB6       =6
.equ        PINB5       =5
.equ        PINB4       =4
.equ        PINB3       =3
.equ        PINB2       =2
.equ        PINB1       =1
.equ        PINB0       =0
.equ        PC7         =7
```

.equ	PC6	=6
.equ	PC5	=5
.equ	PC4	=4
.equ	PC3	=3
.equ	PC2	=2
.equ	PC1	=1
.equ	PC0	=0
.equ	DDC7	=7
.equ	DDC6	=6
.equ	DDC5	=5
.equ	DDC4	=4
.equ	DDC3	=3
.equ	DDC2	=2
.equ	DDC1	=1
.equ	DDC0	=0
.equ	PINC7	=7
.equ	PINC6	=6
.equ	PINC5	=5
.equ	PINC4	=4
.equ	PINC3	=3
.equ	PINC2	=2
.equ	PINC1	=1
.equ	PINC0	=0
.equ	PD6	=6
.equ	PD5	=5
.equ	PD4	=4
.equ	PD3	=3
.equ	PD2	=2
.equ	PD1	=1
.equ	PD0	=0
.equ	DDD6	=6
.equ	DDD5	=5
.equ	DDD4	=4
.equ	DDD3	=3
.equ	DDD2	=2
.equ	DDD1	=1
.equ	DDD0	=0
.equ	PIND6	=6
.equ	PIND5	=5
.equ	PIND4	=4
.equ	PIND3	=3
.equ	PIND2	=2
.equ	PIND1	=1
.equ	PIND0	=0
.equ	SPIF	=7
.equ	WCOL	=6
.equ	SPIE	=7
.equ	SPE	=6
.equ	DORD	=5
.equ	MSTR	=4

```
.equ       CPOL      =3
.equ       CPHA      =2
.equ       SPR1      =1
.equ       SPR0      =0
.equ       RXC       =7
.equ       TXC       =6
.equ       UDRE      =5
.equ       FE        =4
.equ       OR        =3
.equ       RXCIE     =7
.equ       TXCIE     =6
.equ       UDRIE     =5
.equ       RXEN      =4
.equ       TXEN      =3
.equ       CHR9      =2
.equ       RXB8      =1
.equ       TXB8      =0
.equ       ACD       =7
.equ       ACO       =5
.equ       ACI       =4
.equ       ACIE      =3
.equ       ACIC      =2
.equ       ACIS1     =1
.equ       ACIS0     =0
.def       XL        =r26
.def       XH        =r27
.def       YL        =r28
.def       YH        =r29
.def       ZL        =r30
.def       ZH        =r31
.equ       RAMEND =$20+$40+$1ff     ;Adjust for registers and I/O
.equ       INT0addr=$001 ;External Interrupt0 Vector Address
.equ       INT1addr=$002 ;External Interrupt1 Vector Address
.equ       ICP1addr=$003 ;Input Capture1 Interrupt Vector Address
.equ       OC1Aaddr=$004 ;Output Compare1A Interrupt Vector Address
.equ       OC1Baddr=$005 ;Output Compare1B Interrupt Vector Address
.equ       OVF1addr=$006 ;Overflow1 Interrupt Vector Address
.equ       OC0addr =$007 ;Output Compare0 Interrupt Vector Address
.equ       OVF0addr=$008 ;Overflow0 Interrupt Vector Address
.equ       SPIaddr =$009 ;SPI Interrupt Vector Address
.equ       URXCaddr=$00a ;UART Receive Complete Interrupt Vector
                          ;Address
.equ       UDREaddr=$00b ;UART Data Register Empty Interrupt Vector
                          ;Address
.equ       UTXCaddr=$00c ;UART Transmit Complete Interrupt Vector
                          ;Address
.equ       ACIaddr =$00d ;Analog Comparator Interrupt Vector Address
```

Appendix E
Some Fundamentals of RS-232

RS-232 is a serial interface using DB25 or DB9 connectors and, in some cases, RJ45 connectors. RS-232 specifies both the electrical and mechanical interfaces.

RS-232 is an EIA/TIA norm that is equivalent to V.24/V.28 of the CCITT. V.24, however, specifies the mechanical interface and V.28 the electrical. Both (RS-232,V24/V28) define the same interface and will hereafter be referred to as RS-232.

RS-232 defines the cable connecting a DTE (data terminal equipment—intelligent to data) and a DCE (data communication equipment—pass-through device) and specifies the connector. It is a single-ended interface with one lead for every signal and a ground reference.

The DB25 specification, which is officially called RS-232C, is the most common one. There is also RS-232D, which is RS-232 on an RJ45 connector. On an IBM-compatible PC, you will often find a DB9 male for serial connections. This is also RS-232 and is officially called EIA/TIA 574.

According to the specs, the maximum distance is 15 m. With special cabling, distances up to 150 m are possible, but cannot be guaranteed.

Figure E-1 explains the voltage used in RS-232. A positive voltage between +3 and +15 V represents a logical 0 or space, whereas a negative voltage between -3 and -15 V represents a logical 1 or mark.

The pinout for RS-232 connectors is not unique because of the distinction between DTE and DCE. A personal computer normally is a DTE, whereas a modem is a DCE.

The pinouts for both RS-232 connectors used in PCs are listed in Tables E-1 and E-2.

To connect a DTE and a DCE, a straight-through cable (1-1, 2-2, 3-3, etc.) is the most convenient cable to use. If you want to connect two DTEs or two DCEs, you need a null-modem (cross-wired) cable. The distinction between DTE and DCE often leads to confusion.

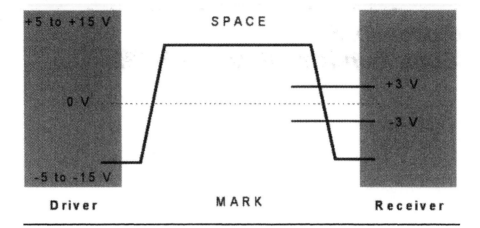

Figure E-1
RS-232 logic level specification.

For a simple data interchange, three wires (Tx, Rx, and GND) are enough. Although the normal PC hardware might well run with just these three wires connected, most driver software will wait forever for one of the handshaking lines to go to the correct level. Depending on the signal state it might work sometimes; other times it might not.

Table E-3
RS-232 pin assignments (DB25 PC signal set).

Pin 1   Protective Ground
Pin 2   Transmit Data
Pin 3   Received Data
Pin 4   Request To Send
Pin 5   Clear To Send
Pin 6   Data Set Ready
Pin 7   Signal Ground
Pin 8   Received Line Signal Detector (Data Carrier Detect)
Pin 20   Data Terminal Ready
Pin 22   Ring Indicator

The connector on the PC has male pins; therefore, the mating cable needs to be terminated in a DB25/F (female pin) connector.

AVR RISC Microcontroller Handbook

Table E-2
RS-232 pin assignments (DB9 PC signal set).

Pin 1   Received Line Signal Detector (Data Carrier Detect)
Pin 2   Received Data
Pin 3   Transmit Data
Pin 4   Data Terminal Ready
Pin 5   Signal Ground
Pin 6   Data Set Ready
Pin 7   Request To Send
Pin 8   Clear To Send
Pin 9   Ring Indicator

The connector on the PC has male pins; therefore, the mating cable needs to be terminated in a DB9/F (female pin) connector.

The reliable solution is to loop back the handshake lines, as shown in Figure E-2, if they are not used. When the lines are handshake-looped, the RTS output from the PC immediately activates the CTS input, so the PC effectively controls its own handshaking.

Connecting together two serial devices (DTE) involves connecting the Rx of one device to the Tx of the other, and vice versa. Figure E-3 indicates how you would go about connecting two PCs together, without handshaking.

After the electrical connection is finished, both sides of the communication channel must understand each other.

To prepare data for transmission from one point to another, most RS-232 systems use dedicated communication controllers. These controllers are mostly called UARTs (universal asynchronous receivers and transmitters) and are responsible for controlling the data exchange over the RS-232 interface.

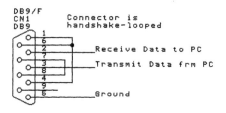

Figure E-2
Handshake loop-back.

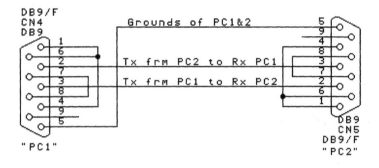

Figure E-3
Connecting two PCs together.

UARTs offload most of the communication activities from the CPU, thus freeing the CPU for other activities. UARTs have the ability to add or remove start and stop bits and provide odd- or even-parity code generation and detection. Currently, components have onboard receiver and transmitter FIFOs for buffering the asynchronous exchanged data.

Figure E-4 shows the structure of the asynchronous transmission protocol used in RS-232 data exchange.

The data transmission starts with a low-level start bit followed by five to eight data bits. An optional parity bit after each character can be generated to check the parity of the received character for error detection on the receiver's side. The transmission of one character ends with the stop bits. The protocol allows 1, 1.5, or 2 stop bits. After the transmission is completed, the line remains at high level until the start bits of the next character return it to low level.

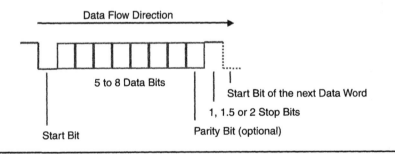

Figure E-4
Asynchronous transmission protocol.

Table E-3
RS-232 parameters.

Baud Rate	9600 bps
Parity	None
Data Bits	8
Stop Bits	1

A further parameter of RS-232 data exchange is the rate of data transmission—the baud rate. The baud rate indicates the number of bits sent or received per second (bps).

A typical configuration for an RS-232 interface in microcontroller applications is shown in Table E-3. With this parameter setup, the transmission time of one character (with 10 bits) is about 1 ms. The parity bit has no function; therefore, no error detection is possible on the receiver's side.

Appendix F
Some Fundamentals of RS-422 and RS-485

RS-422 and RS-485 are other specifications for types of physical connections between serial devices. The actual data transmitted and the voltage transitions used to transmit the individual bytes follow the same pattern as the RS-232 interface that is the standard for PCs.

In contrast to RS-232, RS-422 and RS-485 are data transmission interfaces that use balanced differential signals. Both interfaces are standardized in the EIA/TIA-422 and EIA/TIA-485 Standards.

The major difference between RS-232 and RS-422/485 is the balanced data transmission. In a balanced differential system, the voltage produced by the driver appears across a pair of signal lines that transmit only one signal. Normally, both signal lines are marked as A and B. The resulting voltage V_{AB} defines the logic level, as Figure F-1 shows.

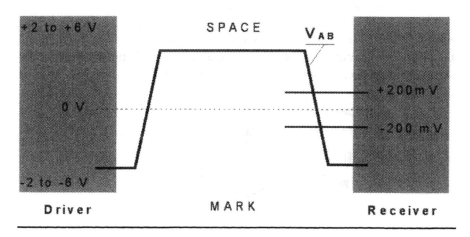

Figure F-1
RS-422/485 Logic level specification.

There is also a difference between RS-422 and RS-485. RS-422 can be compared with RS-232, except for the use of balanced differential signals. The two-wire interface of RS-232 (TxD and RxD) grows up to a four-wire interface, with two lines for the transmit channel and two lines for the receive channel. The RS-422 serial interface works full-duplex as RS-232, meaning that data transmission takes place in both directions at the same time. There is always only one transmitter, but there can be more than one (up to 10) receiver(s).

RS-485 is designed to connect multiple devices with a single twisted-pair connection. Therefore, RS-485 allows half-duplex only: data transmission takes place in only one direction at a time. Up to 32 transceivers (i.e., transmitters and receivers) can be connected.

Nearly each manufacturer of semiconductors offers its own transceiver devices for RS-422 and RS-485. Figure F-2 shows the logic diagram of a typical transceiver device. On the right, the markers A and B characterize both balanced differential serial bus lines.

For RS-485, the driver with its symmetrical output is connected to the symmetrical input of the receiver in the internal transceiver device. For RS-422, the two symmetrical output lines of the transmitter and the two symmetrical input lines of the receiver are each connected to pins of the device. Thus, the four-wire connection between different transceiver devices can be built externally.

As mentioned earlier, data transmission in RS-485 is half-duplex, which means that either the transmitter or the receiver is active. The lines DE (Driver Enable) and /RE (Receiver Enable) control the driver and receiver and can be connected together for RS-485 data transmission.

Figure F-3 shows some RS-485 transceiver devices building a simple network. The line termination resistors on both ends of the data transmission line are very important for reliable data transmission.

When a particular node of the network is not transmitting, its transmitter must be disabled. In an RS-232 to RS-485 converter or an RS-485 serial card, this may be implemented using the RTS control signal from an asynchronous serial port to enable/disable the RS-485 transmitter.

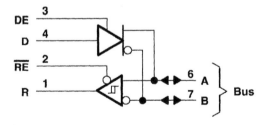

Figure F-2
Transceiver device connected for RS-485.

AVR RISC Microcontroller Handbook

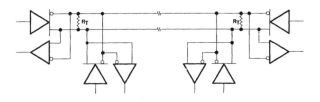

Figure F-3
RS-485 transceivers build a network.

The RTS line can be connected to the RS-485 control lines DE and /RE such that setting the RTS line to Hi enables the RS-485 driver. Setting the RTS line Lo disables the driver by putting it into the tristate condition. This in effect disconnects the driver from the bus, allowing other nodes to transmit over the same wire pair.

When RTS control is used, it is important to be certain that RTS is set high before data is sent. Also, the RTS line must then be set low after the last data bit is sent. This timing is done by the software used to control the serial port and not by the converter.

More details about RS-422 and RS-485 data transmission are explained in an application note from B&B Electronics Mfg. Co. Inc. (see Appendix J).

To close this section, some general aspects should be discussed.

RS-422 and RS-485 are only hardware specifications. Software protocols are not discussed in either specification. It is up to the system designer to define a protocol suitable for the system in question.

A master–slave system has one node that issues commands to each of the "slave" nodes and processes responses. Slave nodes will not typically transmit data without a request from the master node, and they do not communicate with each other. Each slave must have a unique address so that it can be addressed independently of other nodes.

Each node in a multimaster RS-485 system can initiate its own transmission, creating the potential for data collisions. This type of system requires the designer to implement a more sophisticated method of error detection, including methods such as line contention detection, acknowledgement of transmissions, and a system for resending corrupted data.

A CAN-bus based system can serve as an example of a multimaster RS-485 system. Formerly developed for applications in automobiles, systems such as DeviceNet or SDS are now used in industrial environments. For more information about the internals of such multimaster RS-485 systems, see specialized literature (the GPCAN reference given in Appendix J, for example).

Appendix G
8-Bit Intel Hex File Format

The 8-bit Intel hex file format is a printable ASCII format consisting of one or more data records followed by an end-of-file record. Each record consists of one line of information. Data records may appear in any order. Address and data values are represented by 2 or 4 hexadecimal digits.

Record Format

:LL AAAA RR DDDD......DDDD CC (spaces included for better readability)

Character	Description
LL	Length field. Number of data bytes.
AAAA	Address field. Address of first byte.
RR	Record type field. 00 for data and 01 for end of record.
DD	Data field.
CC	Checksum field. One's complement of length, address, record type, and data fields modulo 256.

Example

:06 0100 00 010203040506 E4 (spaces included for better readability)

:00 0000 01 FF (spaces included for better readability)

The first line in this example Intel hex file is a data record addressed at location 100H with data values 1 to 6. The second line is the end-of-file record.

Appendix H
Decimal-to-Hex-to-ASCII Converter

The following conversion table will help in converting between different notations of ASCII data.

For hexadecimal numbers, two notations are given, the more C-like notation 0x40 for the character "@" and the more BASIC-like notation $40. Both notations are equivalent and mean the same.

For control characters (value < 0x20) the required keystrokes for input are listed. All other characters can be input via keystrokes. Uppercase and lowercase are distinguished only by the shift key.

DEC	HEX	HEX	ASCII	Key	DEC	HEX	HEX	ASCII	Key
0	0x00	$00	NUL	^@	64	0x40	$40	@	
1	0x01	$01	SOH	^A	65	0x41	$41	A	
2	0x02	$02	STX	^B	66	0x42	$42	B	
3	0x03	$03	ETX	^C	67	0x43	$43	C	
4	0x04	$04	EOT	^D	68	0x44	$44	D	
5	0x05	$05	ENQ	^E	69	0x45	$45	E	
6	0x06	$06	ACK	^F	70	0x46	$46	F	
7	0x07	$07	BEL	^G	71	0x47	$47	G	
8	0x08	$08	BS	^H	72	0x48	$48	H	
9	0x09	$09	HT	^I	73	0x49	$49	I	
10	0x0A	$0A	LF	^J	74	0x4A	$4A	J	
11	0x0B	$0B	VT	^K	75	0x4B	$4B	K	
12	0x0C	$0C	FF	^L	76	0x4C	$4C	L	
13	0x0D	$0D	CR	^M	77	0x4D	$4D	M	
14	0x0E	$0E	SO	^N	78	0x4E	$4E	N	
15	0x0F	$0F	SI	^O	79	0x4F	$4F	O	
16	0x10	$10	DLE	^P	80	0x50	$50	P	
17	0x11	$11	DC1	^Q	81	0x51	$51	Q	
18	0x12	$12	DC2	^R	82	0x52	$52	R	
19	0x13	$13	DC3	^S	83	0x53	$53	S	
20	0x14	$14	DC4	^T	84	0x54	$54	T	
21	0x15	$15	NAK	^U	85	0x55	$55	U	
22	0x16	$16	SYN	^V	86	0x56	$56	V	

DEC	HEX	HEX	ASCII	Key	DEC	HEX	HEX	ASCII	Key
23	0x17	$17	ETB	^W	87	0x57	$57	W	
24	0x18	$18	CAN	^X	88	0x58	$58	X	
25	0x19	$19	EM	^Y	89	0x59	$59	Y	
26	0x1A	$1A	SUB	^Z	90	0x5A	$5A	Z	
27	0x1B	$1B	ESC	^[	91	0x5B	$5B	[	
28	0x1C	$1C	FS	^\	92	0x5C	$5C	\	
29	0x1D	$1D	GS	^]	93	0x5D	$5D	]	
30	0x1E	$1E	RS	^^	94	0x5E	$5E	^	
31	0x1F	$1F	US	^_	95	0x5F	$5F	_	

DEC	HEX	HEX	ASCII	Key	DEC	HEX	HEX	ASCII	Key
32	0x20	$20	SP		96	0x60	$60	`	
33	0x21	$21	!		97	0x61	$61	a	
34	0x22	$22	"		98	0x62	$62	b	
35	0x23	$23	#		99	0x63	$63	c	
36	0x24	$24	$		100	0x64	$64	d	
37	0x25	$25	%		101	0x65	$65	e	
38	0x26	$26	&		102	0x66	$66	f	
39	0x27	$27	'		103	0x67	$67	g	
40	0x28	$28	(		104	0x68	$68	h	
41	0x29	$29	)		105	0x69	$69	i	
42	0x2A	$2A	*		106	0x6A	$6A	j	
43	0x2B	$2B	+		107	0x6B	$6B	k	
44	0x2C	$2C	,		108	0x6C	$6C	l	
45	0x2D	$2D	-		109	0x6D	$6D	m	
46	0x2E	$2E	.		110	0x6E	$6E	n	
47	0x2F	$2F	/		111	0x6F	$6F	o	
48	0x30	$30	0		112	0x70	$70	p	
49	0x31	$31	1		113	0x71	$71	q	
50	0x32	$32	2		114	0x72	$72	r	
51	0x33	$33	3		115	0x73	$73	s	
52	0x34	$34	4		116	0x74	$74	t	
53	0x35	$35	5		117	0x75	$75	u	
54	0x36	$36	6		118	0x76	$76	v	
55	0x37	$37	7		119	0x77	$77	w	
56	0x38	$38	8		120	0x78	$78	x	
57	0x39	$39	9		121	0x79	$79	y	
58	0x3A	$3A	:		122	0x7A	$7A	z	
59	0x3B	$3B	;		123	0x7B	$7B	{	
60	0x3C	$3C	<		124	0x7C	$7C			
61	0x3D	$3D	=		125	0x7D	$7D	}	
62	0x3E	$3E	>		126	0x7E	$7E	~	
63	0x3F	$3F	?		127	0x7F	$7F	DEL	

AVR RISC Microcontroller Handbook

Appendix I
Overview of Atmel's Application Notes and Software

Atmel publishes application notes and software on its Web sites. The URL for the application notes is http://www.atmel.com/atmel/products/prod201.htm. Software can be found under http://www.atmel.com/atmel/products/prod 203.htm.

AVR RISC—Application Notes As of September 22, 1997, the following AVR RISC application notes were available.

AVR000: Register and Bit-Name Definitions for the AVR Microcontroller (1 page) This application note contains files that allow the use of register and bit names from the databook when writing Assembly programs. See the Software section to download avr000.exe.

AVR100: Accessing the AT90S1200 EEPROM (5 pages) This application note contains routines for access of the EEPROM memory in the AT90S1200. See the Software section to download avr100.asm.

AVR102: Block Routines (3 pages) This application note contains routines for transfer of data blocks. See the Software section to download avr102.asm.

AVR128: Setup and Use of the Analog Comparator (3 pages) This application note serves as an example of how to set up and use the AVR's on-chip analog comparator. See the Software section to download avr128.asm.

AVR190: Power-Up Considerations (4 pages) This application note describes how to ensure proper operation of an AVR Microcontroller after power-up.

AVR200: Multiply and Divide Routines (19 pages) This application note lists subroutines for multiplication and division of 8- and 16-bit signed and unsigned numbers. See the Software section to download avr200.exe.

AVR202: 16-Bit Arithmetics (2 pages)   This application note lists program examples for arithmetic operation on 16-bit values. See the Software section to download `avr202.asm`.

AVR204: BCD Arithmetics (11 pages)   This application note lists routines for BCD arithmetics. See the Software section to download `avr204.asm`.

AVR220: Bubble Sort (3 pages)   This application note implements the Bubble Sort algorithm on AVR controllers. See the Software section to download `avr220.asm`.

AVR222: 8-Point Moving Average Filter (3 pages)   This application note demonstrates how the addressing modes in the AVR architecture can be utilized. See the Software section to download `avr222.asm`.

AVR304: Half Duplex Interrupt Driven Software UART (11 pages)   This application note describes how to make a half-duplex UART on any AVR device using the 8-bit Timer/Counter0 and an external interrupt. See the Software section to download `avr304.asm`.

AVR400: Low Cost A/D Converter (5 pages)   This application note targets cost- and space-critical applications that need an ADC. See the Software section to download `avr400.asm`.

AVR910: In-System Programming (9 pages)   This application note shows how to design the system to support in-system programming.

AVR RISC—Software   As of September 22, 1997, the following AVR RISC software was available.

AVR.EXE (1244 kb)   Self-extracting archive containing AT90S (AVR) Family Assembler and Simulator software version 1.11. User manuals are included in Adobe Acrobat format.

ASTUDIO.EXE (1481 kb)   Self-extracting archive containing AVR Studio version 1.20, a C and Assembler source-level debugger for the AT90S (AVR) family. Requires Windows95 or Windows NT.

APROGWIN.EXE (859 kb)   Self-extracting archive containing AVR Prog Version 1.21, Windows-based software for using the AVR Development Board to program AVR Devices. Requires Windows95 or Windows NT.

DB_UG_71.ZIP (5 kb) Archive containing software update for the AT90S1200DEV AVR development board. This software update allows the development board to support serial programming of all AT90S8515 parts.

AVR000.EXE (24 kb) Self-extracting archive containing 1200DEF.INC, 2313DEF.INC, 4414DEF.INC, and 8515DEF.INC for AVR000: Register and Bit-Name Definitions for the AVR MCU Application Note.

AVR100.ASM (6 kb) Code for AVR100: Accessing the AT90S1200 EPROM Application Note.

AVR102.ASM (4 kb) Code for AVR102: Block Copy Routines Application Note.

AVR128.ASM (6 kb) Code for AVR128: Set up and Use the Analog Comparator Application Note.

AVR200.EXE (23 kb) Self-extracting archive containing AVR200.ASM and AVR200B.ASM for AVR200: Multiply and Divide Routines Application Note.

AVR202.ASM (8 kb) Code for AVR202: 16-Bit Aritimetics Application Note.

AVR204.ASM (13 kb) Code for AVR204: BCD Aritimetics Application Note.

AVR220.ASM (4 kb) Code for AVR220: Bubble Sort Application Note.

AVR222.ASM (5 kb) Code for AVR222: 8-Point Moving Average Filter Application Note.

AVR304.ASM (13 kb) Code for AVR304: Half Duplex Interrupt Driven Software UART Application Note.

AVR400.ASM (8 kb) Code for AVR400: Low Cost A/D Converter Application Note.

Appendix J
Literature

Campbell, D.: "Designing for Electromagnetic Compatibility with Single-Chip Microcontrollers." Application Note AN1263, Motorola, Inc. (1995).

GPCAN—a general-purpose CAN-bus node. "Hardware Overview and Applications." http://www.nikhef.nl/pub/departments/ct/po/doc/gpcan-html/gpcan.html

RS-422 and RS-485 Application Note, B&B Electronics Mfg. Co. Inc., http://www.bb-elec.com/bb-elec/literature/485appnote.pdf

Appendix K
Contacts

☒ Atmel Corp.
Corporate Headquarters
2325 Orchard Parkway
San Jose, CA 95131
☎ (408) 441-0311
🖷 (408) 436-4300

☒ Atmel Colorado Springs
1150 E. Cheyenne Mountain Blvd.
Colorado Springs, CO 80906
☎ (719) 576-3300
🖷 (719) 540-1759
💻 http://www.atmel.com/

☒ IAR Systems Inc.
One Maritime Plaza
San Francisco,
CA 94111
☎ (415) 765-5500
🖷 (415) 765-5503
💻 info@iar.com

☒ IAR Systems AB
P.O. Box 23051
S-750 23 Uppsala,
Sweden
☎ +46-18-16-78-00
🖷 +46-18-16-78-38
💻 info@iar.se
💻 info@iarsys.co.uk
http://www.iar.se

☒ IAR Systems Ltd.
9 Spice Court,
Ivory Square
London SW11 3UE,
England
☎ +44-171-924-3334
🖷 +44-171-924-5341

☒ Equinox Technologies   *Programmer and Evaluation Boards*
229 Greenmount Lane, Bolton BL1 5JB UK
☎ +44-1204-492010
🖷 +44-1204-494883
💻 sales@equinox-tech.com
http://www.equinox-tech.com

☒ E-LAB Computers   *Pascal Compiler*
Grombacher Strasse 27
D-74906 Bad Rappenau, Germany
☎ +49-7268-9124-0
🖷 +49-7268-9124-24

💻 http://www.sistudio.com/avr.html   *SiStudio AVR Resources*
http://www.avrbasic.com/appnotes/   *SiStudio AVR BASIC Application Notes*

✉ DonTronics   *Microcontroller Kits and Components*
P.O. Box 595, Tullamarine 3043, Australia
(29 Ellesmere Cres., Tullamarine 3043, Australia)
☎ +613-9338-6286
🖩 +613-9338-2935
💻 http://dontronics.com

✉ Dr. Claus Kühnel   *Microcontroller Applications and*
Schlyffistr. 14, CH-8806 Bäch   *Download of Program Examples*
Switzerland
☎ +41-1-7850238
🖩 +41-1-7850275
💻 ckuehnel@access.ch
http://www.access.ch/ckuehnel

Index

Printed and bound by CPI Group (UK) Ltd, Croydon, CR0 4YY

03/10/2024

01040432-0011